# Fundamentals of Digital Imaging in Medicine

Roger Bourne

# Fundamentals of Digital Imaging in Medicine

 Springer

Roger Bourne, PhD
Discipline of Medical Radiation Sciences
Faculty of Health Sciences
University of Sydney
Sydney
Australia

Additional material to this book can be downloaded from http://extra.springer.com.

ISBN 978-1-84882-086-9          ISBN 978-1-84882-087-6 (eBook)
DOI 10.1007/978-1-84882-087-6
Springer London Dordrecht Heidelberg New York

British Library Cataloguing in Publication Data
A catalogue record for this book is available from the British Library

Library of Congress Control Number: 2009929390

*Cover design*: eStudio Calamar S.L.

Printed on acid-free paper

Springer is part of Springer Science+Business Media (www.springer.com)

For Joan and John,
Who gave me curiosity and scepticism.

*Whit dae birds write on the dusk?*
*A word niver spoken or read,*
*The skeins turn hame,*
*on the wind's dumb moan, a soun,*
*maybe human, bereft.*

*Kathleen Jamie*

# Foreword

There was a time not so long ago, well within the memory of many of us, when medical imaging was an analog process in which X-rays, or reflected ultrasound signals, exiting from a patient were intercepted by a detector, and their intensity depicted as bright spots on a fluorescent screen or dark areas in a photographic film. The linkage between the exiting radiation and the resulting image was direct, and the process of forming the image was easily understandable and controllable. Teaching this process was straightforward, and learning how the process worked was relatively easy. In the 1960s, digital computers began to migrate slowly into medical imaging, but the transforming event was the introduction of X-ray computed tomography (CT) into medical imaging in the early 1970s. With CT, the process of detecting radiation exiting from the patient was separated from the process of forming and displaying an image by a multitude of computations that only a computer could manage. The computations were guided by mathematical algorithms that reconstructed X-ray images from a large number of X-ray measurements across multiple imaging planes (projections) obtained at many different angles. X-ray CT not only provided entirely new ways to visualize human anatomy; it also presaged the introduction of digital imaging methods to every imaging technique employed in medicine, and ushered the way for new imaging technologies such as magnetic resonance and optical imaging. Digital imaging permits image manipulations such as edge enhancement, contrast improvement and noise suppression, facilitates temporal and energy subtraction of images, and speeds the development of hybrid imaging systems in which two (or more) imaging methods can be deployed on the same gantry and without moving the patient. The production and manipulation of digital images are referred to collectively as imaging processing.

Without question, the separation of signal detection from image display offers many advantages, including the ability to optimize each process independently of the other. However, it also presents a major difficulty, namely that to many persons involved in imaging, the computational processes between detection and display are mysterious operations that are the province of physicists and engineers. Physicians, technologists and radiological science students are expected to accept the validity of the images produced by a mysterious 'black box' between signal input and image output without really understanding how the images are formed from input signals.

A plethora of text and reference books, review articles and scientific manuscripts have been written to describe the mechanisms and applications of the various mathematical algorithms that are used in image processing. These references are interpretable by the mathematical cognoscenti, but are of little help to most persons who lack the mathematical sophistication of physicists and engineers. What is needed is a text that explains image processing without advanced mathematics so that the reader can gain an intuitive feel for what occurs between signal detection and image display. Such a text would be a great help to many who want to understand how images are formed, manipulated and displayed but who do not have the background needed to understand the mathematical algorithms used in this process.

Roger Bourne has produced such a text, and he will win many friends through his efforts. The book begins with a brief description of digital and medical images, and quickly gets to what I believe is the most important chapter in the book: Chapter 4 on Spatial and Frequency Domains. This chapter distinguishes between spatial and frequency domains, and then guides the reader through Fourier transforms between the two in an intuitive and insightful manner and without complex mathematics. The reader should spend whatever time is needed to fully comprehend this chapter, as it is pivotal to understanding digital image formation in a number of imaging technologies. Following a discussion of Image Quality, the reader is introduced to various image manipulations for adjusting contrast and filtering different frequencies to yield images with heightened edges and reduced noise. Chapter 7 on Image Filters is especially important because it reveals the power of working in the frequency domain permitted by the Fourier process. After an excellent chapter on Spatial Transformation, the author concludes with four appendices, including a helpful discussion of ImageJ, a software package in the public domain that is widely used in image processing. This discussion provides illustrations of a powerful tool for image manipulation.

Altogether too often we in medical imaging become enamored with our technologies and caught up in the latest advances replete with jargon, mathematics, and other arcane processes. We forget what it was like when we entered the discipline, and today the discipline is far more complex than it was even a few short years ago. That is why a book such as Dr. Bourne's is such a delight. This book guides the reader in an intuitive and common sense manner without relying on sophisticated mathematics and esoteric jargon. The result is a real 'feel' for image processing that will serve the reader well into the future. We need more books like it.

Milwaukee, Wisconsin                                                   *William Hendee*
March 30, 2009

# Preface

Do we really need *another* digital imaging text? What, if anything, is special about this one? The students I teach, medical radiation science undergraduates, have said *'Yes we do'*. The rapid movement of medical imaging into digital technology requires graduates in the medical radiation sciences to have a sound understanding of the fundamentals of digital imaging theory and image processing — areas that were formerly the preserve of engineers and computer scientists. There are many excellent texts written for the mathematically adept and well trained, but very few for the average radiation science undergraduate who has only high school maths training. This book is for the latter.

Some notable features of this book are:

- Scope: It focuses on medical imaging.
- Approach: The approach is intuitive rather than mathematical.
- Emphasis: The concept of spatial frequency is the core of the text.
- Practice: Most of the concepts and methods described can be demonstrated and practiced with the free public-domain software ImageJ.
- Revision: Major parts can be revised by studying just the figures and their captions.

Radiographers, radiation therapists, and nuclear medicine technologists routinely acquire, process, transmit and store images using methods and systems developed by engineers and computer scientists. Mostly they *don't* need to understand the details of the maths involved. However, everyone does their job better, and has a better chance of improving the way their job is done, when they understand the tools they use at the deepest possible level. This book tries to dig as deep as possible into imaging theory without using maths.

I have aimed to describe the basic properties of digital images and how they are used and processed in medical imaging. No realistic discussion of image manipulation, and in the case of MRI, image formation, can escape the bogey man, Joseph Fourier. One of the novelties of this text is that it cuts straight to the chase and starts with the concept of spatial frequency. I have attempted to introduce this concept in a purely intuitive way that requires no more maths than a cosine and *the idea* of a complex number. The mathematically inclined may think my explanation takes a very long path around a rather small hill. I hope the intended audience will be glad of the detour. Expressions for the Cosine, Hartley, and Fourier transforms are included more as pictures than as tools. I believe it is possible for my readers to get an understanding of what the transforms do without being able, nor ever needing, to implement them from first principles.

A second novelty of the text is the images and illustrations. Many of these are synthetic (thanks mostly to MatLab) because I believe it is easier to understand a concept when not distracted by irrelevant information. The images start simple and get more complicated as the level of discussion deepens. When a concept or method has been explored with simple images I try to provide illustrations using real medical images. To some extent the captions for illustrations repeat explanations present in the text. Apart from the learning value of repetition I have done this in an attempt to make the images and their captions self-explanatory. My intention is that the reader will be able to revise the major chapters of the text simply by studying the illustrations and their captions.

Many of the principles and techniques described can be practically explored using the public domain image processing software ImageJ. ImageJ is not a toy. It is used worldwide in medical image processing, especially in research, and the user community is continuously developing new problem-specific tools which are made available as plugins. An introduction to ImageJ is thus likely to be of long-term benefit to a medical radiation scientist. Where appropriate the text includes reference to the relevant ImageJ command or tool, and many illustrations show an ImageJ tool or output window. A very brief introduction to ImageJ is included as an Appendix, however, this text is in no way an ImageJ manual.

Perhaps it is appropriate to justify the omission of two major topics – image analysis and image registration. These are important tools vital to modern medical imaging. However, they are both large and complex fields and I could not envisage a satisfactory, non-trivial, way to introduce them in a text that is a *primer*. If I am told this is a major omission then I will address the problem in a second edition. For now, I hope that this text's focus on the basic principles of digital imaging gives students a solid intuitive foundation that will make any later encounters with image analysis and registration more comfortable and productive.

To all the people who have helped me in various ways with the development and writing of this book, whether through suggestions, or simple tolerance, I give my warm thanks – especially Toni Shurmer, Philip Kuchel, Chris Constable, Terry Jones, Jane and Vickie Saye, Jenny Cox, and Roger Fulton. It has been a task far bigger than I anticipated but nevertheless a rewarding and educational one. My daughters will be interested to see *that book* as a physical object, though it's probably not one they would willingly choose to investigate. My parents will be pleased to see I have done something besides fall off cliffs. I extend particular thanks to the staff at Springer who have been very patient, and I am deeply honored by Bill Hendee's foreword. Not least, I thank my past students for their feedback and tolerance in having to test drive many even more imperfect versions than the one you hold now. If they ran off the road I hope their injuries were minor.

Despite a large amount of 'iterative reconstruction' I don't pretend this text is ideal in content, detail, fact, or approach. I look forward to comments and suggestions from students, academics, and practitioners on how it can or might be improved. Please email me: rbourne@usyd.edu.au.

The manuscript for this text was prepared with TeXnicCenter and MiKTeX – a Windows PC based integrated development environment for the LaTeX typesetting language (www.texniccenter.org). This software has been a pleasure to use and the developers are to be commended for making it freely available to the public.


Sydney                                                            *Roger Bourne*
December, 2009

# Contents

# Chapter 1
# Introduction

The universe is full of spinning objects – galaxies, suns, planets, weather patterns, pink ballerinas, footballs, atoms, and subatomic particles to name a few. It is remarkable *not* that humans invented the wheel, but that they took *so long*. Bacteria did it millions of years earlier. However, humans are remarkable for their powers of observation, virtual memory (recording), and analysis. The wheel of the mind, a much more remarkable invention than the wheel of the donkey cart or the Ferrari, is mathematics. Just as recording extends human memory beyond its physical limitations, mathematics extends human analysis into regions inconceivable to the mind – complex numbers being a particularly apposite example. If you use mathematics to describe the appearance of a spinning object the answer is a sinusoid. If you use mathematics to describe the behavior of the energy used for medical imaging the answer is a sinusoid. In MRI the spinning object and the energy used for imaging are inseparable. Joseph Fourier showed we can go even further than this – every measurable thing, including medical images, can be described with sinusoids. This simple concept, once apprehended, can be seen to bind the multiplicity of medical imaging methods into one whole.

## 1.1 What Is This Book Trying To Do?

Those new to imaging science, and especially those without a background in the mathematical or physical sciences, often find the 'science' of image processing texts bewilderingly mathematical and inaccessible. Yet the majority of technologists that acquire and process medical images do not need to understand the mathematics involved. Few pilots are experts in either engineering or theoretical aerodynamics, yet without a basic understanding of both they can neither qualify nor work. This is reassuring for airline passengers. Similarly, medical technology graduates should be expected to understand the basics of imaging theory and image processing before they practice. This *primer* aims to provide a *working knowledge* of digital imaging theory as used in medicine, *not* a mathematical foundation. With that understanding I hope that the reader could, if curious or required, be able to delve into the more mathematical texts and research papers with a feeling of familiarity

R. Bourne, *Fundamentals of Digital Imaging in Medicine*,
DOI 10.1007/978-1-84882-087-6_1, © Springer-Verlag London Limited 2010

and basic competence. The mathematics may well remain intimidating, or even incomprehensible, but its purpose will hopefully be clear.

The approach of this text is intended to be 'holistic', by which I mean that I have tried to develop and emphasize a core of imaging science theory specific to medical imaging. This is quite deliberately done at the expense of detail and coverage. Specific examples are included because they illustrate a principle, not because they are considered essential or more important than other methods. The concept of spatial frequency and Fourier transforms is introduced early – as soon as the basic characteristics of digital images are explained. Many of the techniques applied directly to spatial domain images have terminology specifically related to the spatial frequency characteristics of the image. It is my intention to demystify these terms as soon as they are introduced in order to minimize both the potential for confusion and the need to ask the reader to wait for an explanation that will come later.

Nearly all of medical imaging is based on making visible light images from measurements of energy that is invisible to humans. Most of the content of this text deals with the principles of 'image data processing', rather than simply 'image processing' – a term many readers would consider included only methods for handling 'constructed' images or postprocessing of images that are the output of medical imaging systems. The issues that need to be considered in handling medical image data include:

- The limitations of the technology used for acquisition
- The characteristics of human visual perception
- The need to simplify or extract specific information from images
- The complex interactions between the above

A quick browse through this book will reveal a number of non-medical and non-human images. There are images of fruit, vegetables, mouse brains, and completely artificial constructs synthesized in my own computer. I am sure most readers will be more than adequately familiar with medical images. I give my readers credit for being able to generalize the points made by use of non-medical images, and to enjoy the beauty of some of the more unusual images. I have used medical images when illustrating some specific feature of medical images.

## 1.2   Chapter Outline

The following notes outline the intended purpose of each chapter in this text.

### 1.2.1   Digital Images

This book is about *digital* image data – including the raw measurement data that is processed to make medical images. The first chapter introduces the idea of storing measurements of imaging energy as discrete arrays: how measurements are represented in digital form; how the storage format affects the precision and potential information content of the stored data; how essential auxiliary and supplementary information is stored; the features of common image file formats; and image data compression.

### 1.2.2 Medical Images

This chapter describes the basic similarity of all medical imaging methods – they all seek to measure *differential flow* of energy through or from the body, the main differences being the location of the energy source. All methods, bar one (ultrasound), measure the flow of photons, and all, including ultrasound, are described or analyzed using wave terminology. The *differences* between the imaging methods are a result of the way the energy interacts with tissue, the way the energy is measured, and the way the measurements are processed to make a visible light image. Different methods give different types of contrast, or the same contrast faster or in more detail.

### 1.2.3 The Spatial and Frequency Domains

This chapter introduces the concept of spatial frequency and takes a very gentle and intuitive path to the 2D Fourier transform. Most of the discussion is about the Fourier spectra of images because this is the most common representation of frequency domain data. However, we also look at the underlying complex data and the meaning of phase which is of particular relevance to MRI.

### 1.2.4 Image Quality

It is one thing to acquire an image but technologists and clinicians who use medical images must be acutely aware of image quality. Without adequate contrast and resolution an image is useless, and both these features are diminished by noise. This chapter looks at methods of description of image quality and imaging system performance – they inevitably include the idea of spatial frequency.

### 1.2.5 Contrast Adjustment

Human visual perception has quite poor and non-linear discrimination of light intensity. For this reason one of the most common image processing adjustments is the selective improvement of contrast. The raw information encoded in small differences of image intensity may be invisible to a human until these differences are exaggerated by contrast adjustment.

The necessity for contrast adjustment also arises from the imaging technology. In the case of a camera the sensor has a response to light intensity which is different from the response of the human eye. In medical imaging the contrast measured is, in general, not even a variation in visible light intensity. Ultrasound, X-ray, magnetic resonance, PET and SPECT imaging are all technologies where a visible light image is used to display measured energy differences that are invisible to humans. The images produced have no 'native' visible light format and thus automatically require some form of contrast adjustment.

### 1.2.6   Image Filters

Filtering of image data is possibly an even more common operation than contrast adjustment though often it occurs before creation of a visible image. This chapter introduces frequency domain filters before spatial domain filters because many of the latter have names that reflect their spatial frequency effects. The equivalence of spatial domain convolution and frequency domain multiplication is emphasized. The focus is on the idea of using a filter to extract or enhance image information, rather than a complete coverage of all commonly used filters.

### 1.2.7   Spatial Transformations

The final chapter looks at the interpolation methods used for spatial transformations of images. Resizing or rotating images means the *available* information in the image has to be used to make a new version of the image. We emphasize that new information cannot be created, though artifacts and distortions can.

### 1.2.8   Appendices

For reference, three appendices that cover important background detail are included: An introduction to get the reader up and running with ImageJ; a clarification of the terms *Precision* and *Accuracy*; and a brief introduction to complex numbers.

## 1.3   Revision

Each chapter concludes with a summary of the most important concepts covered. I suggest that in reviewing the text a reader first rereads the summary items. If the ideas behind a particular item are not fully clear then the relevant section should be studied again.

The second suggested method of review is to work through the figures and their captions. Important concepts from the text are repeated in the figure captions with the intention of making the figures as self-explanatory as possible.

## 1.4   Practical Image Processing

Students will invariably find their grasp of imaging theory improves with some actual practice of image processing. While most commonly available image processing software (commercial and freeware) will enable practice of simple tasks such as display of histograms and contrast adjustment, few stray outside the spatial domain. Most are designed for processing color photographs, not medical images. I therefore recommend that readers download and use the Java-based tool ImageJ from the US National Institute of Health website (details in Appendix A). ImageJ is used extensively worldwide and an active user community is constantly developing new task-specific tools (plugins) which can be installed into the base version as macros. To reduce the potential for confusion I have endeavored to keep the nomenclature used in the text consistent with that used in ImageJ.

### 1.4.1   Images for Teaching

The illustrations used in this text are available on the included CD.

# Chapter 2
# Digital Images

## 2.1 Introduction

What is a digital image? Interestingly this question does not have a simple answer. Consider, for example, this image of a familiar Australian landmark (Fig. 2.1).

**Fig. 2.1** Is this a digital image? No, it's an ink image. The intensity data was stored and manipulated in digital format between the time of capture and the time of printing of this page. Was a digital camera used? There is no way to tell from the ink in this image

What does it mean if we say this is a digital image? The image is printed on the page with ink so there is nothing 'digital' in what we see when we look at the image on the page. Even if the resolution were so poor that we could see pixelation we would not be seeing actual pixels (the smallest elements of image information) but a representation of them. There were many steps between the capture of the visible light image and the printing of the image on this page. It was originally captured with a digital camera, which means the continuous pattern of light being reflected off the Sydney opera house and the harbor bridge was initially recorded as an array of electric charges on a semiconductor light sensor. The amount of charge on each element of the sensor was then measured, converted into a binary number, copied into the memory of the camera, processed in some way, and then written onto a

R. Bourne, *Fundamentals of Digital Imaging in Medicine*,
DOI 10.1007/978-1-84882-087-6_2, © Springer-Verlag London Limited 2010

compact flash card. Later the image was downloaded from a card onto the hard disk in a computer, processed with some software, then stored again in a different format on a hard disk. It would be a very long and tedious story if we traced the path of the image all the way to this printed page. The point to consider is that, at almost every step of this process, the image data would have been stored on different electronic or optical media in different ways. Thus a single image data set can have many different physical forms and we only actually *see* the image when it is converted to a physical form that reflects, absorbs, or emits visible light.

We might broadly separate images into two categories – the measured, and the synthetic. A measured image is one acquired by using some device or apparatus to measure a signal coming from an object or a region of space. Obvious examples are photographs, X-ray images, magnetic resonance images, etc. In contrast, a synthetic image is one not based on a measured signal but constructed or drawn. Typical examples are diagrams, paintings, and drawings. Of course these two broad categories overlap to some extent. Many synthetic images are based on what we see, and many measured images are manipulated to change how we see them and to add further information – lines, arrows, labels, etc.

## 2.2   Defining a Digital Image

In a digital camera the subject light 'pattern' is focused by the lens onto a flat rectangular photosensor and recorded as a rectangular array of picture elements – *pixels*. In the sensor a matrix of photosites accumulate an amount of charge that (up to the saturation point) is proportional to the number of incident photons – the intensity of the light multiplied by the duration of the exposure.

Just how different is this 'digital' process from the so called 'analog' photochemical film process? Not very. With a film camera the subject light pattern is recorded as an *irregular* matrix of silver granules, the film grain, embedded in a thin layer of gelatin. Development of a film image is the chemical process of converting light-activated silver halide grains to an emulsion of silver metal with stable light reflection and transmission properties. By analogy, 'development' of a digital camera image is the process of converting the charge stored on the semiconductor light sensor to a binary array stored on stable electronic media. The stored digital image data is then equivalent to a film negative – it is the stable raw data from which a visible image can be repeatedly produced. Since this happens automatically inside the camera it is not something we pay much attention to.

Whether image contrast is stored as an irregular array, as in film, or a regular array, as in a digital recording, is of no significance in determining the information content (Fig. 2.2). However, it is *much, much* easier to copy, analyze, and process a digital data array.

One of the main operational differences between digital and film sensors is that digital sensors are relatively linear in their response to light over a wide range of exposures while films are generally linear only over a narrow range of exposures. This makes film harder to use because there is much more potential for exposure

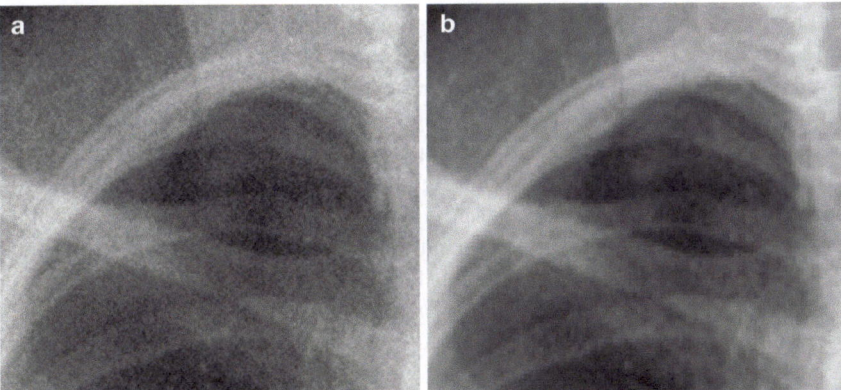

**Fig. 2.2** Illustration of the *lack of* difference between the way film and a direct digital sensor record image information. Image **a** represents the *random* array of silver granules that provide optical contrast in a film recording of image data. Image **b** represents the *rectangular* array of pixel intensities (converted to some display medium) that provide optical contrast in a direct digital recording. There is no significant difference in the *information content* of the two images

errors that lead to either inadequate or excessive film density in the developed image. On the other hand the large dynamic range of digital X-ray detectors means that high exposures still give good quality images. This has led to 'exposure creep' – a gradual increase in routine exposures and unnecessarily high patient doses.

Another, less direct, analog of the chemical process of film development is the process of *image reconstruction*. Image reconstruction is the term used to describe the methods of formation of anatomical images from the raw data acquired in tomographic (cross-sectional) medical imaging devices. Since the raw data is not a cross-sectional image the process might be more appropriately named *image construction*, however, we will stick to the common usage in this text. Either way, image (re)construction depends on the processing of raw digital data to create a 2D or 3D image in which the position of objects in the image correspond to their positions in the subject – they are not superimposed as in a projection image.

Where does this leave us in defining a digital image? As a working definition we might simply say that *a digital image is an encoding of an image amenable to electronic storage, manipulation and transmission*. This is the huge advantage of digital images over film images. There are numerous ways to do the encoding, manipulation and transmission, each method having specific advantages and disadvantages depending on the intended use of the image. We will definitely not discuss these methods comprehensively, nor in detail, but important points of relevance to medical images will be covered.

No matter how a digital image is stored or handled inside a computer it is displayed as a rectangular array (or matrix) of independent pixels. Of course the objects we image are not rectangular arrays of homogeneous separate elements. The original *continuous* pattern of signal intensity coming from the imaged object is converted by the imaging system into a rectangular array of intensities by *discrete* sampling. Each element of the rectangular array represents the average signal intensity in a small region of the original continuous signal pattern. The size of each small region from which the signal is averaged is determined by the geometry of the imaging system and the physical size of each sensor element.

It is important to remember that the signals from separate regions of the imaged object are not perfectly separated and separately measured by an imaging system. All imaging devices 'blur' the input signal to a certain extent so that the signal recorded for each discrete pixel that nominally represents a specific region of sample space always contains some contribution from the adjacent regions of sample space. This inevitable uncertainty about the precise spatial origin of the measured signal can be described by the *Point Spread Function* (PSF) – an important tool in determining the spatial resolution of an imaging system. The PSF describes the shape and finite size of the small 'blob' we would see if we imaged an infinitely small point source of signal.

The raw image data has a specific size – $m$ pixels high by $n$ pixels wide. Put another way, the image matrix has $m$ rows and $n$ columns. In many image formats the pixel data is not actually *stored* as an $m \times n$ rectangular array. Because most images have large areas of identical or very similar pixels it is often more space and time efficient to store and transmit the pixel information in some compressed form rather than as the full $m \times n$ array. An image stored in this way must be converted back into an $m \times n$ matrix before display.

So far we have discussed only 2D images. In many imaging modalities it is common to construct 3D or *volume* images – effectively a stack of 2D images or slices. This does not change our conception of a digital image – 2D or 3D, it is still a discrete sampling where each pixel or *voxel* (volume element) represents a measurement of the average signal intensity from a region in space.

When we open a digital image file the computer creates a temporary $m \times n$ array of pixel data based on the information in the image file (if it is a color image then a series of $m \times n$ arrays are created – one for each base color, e.g. red, green, and blue in the case of an RGB image). This array is the one on which any image processing is performed, or it provides the input data for image processing that outputs a new 'processed image' array. If the image is to be displayed on a computer monitor then the rectangular array of pixel intensity and color information is converted into a new array that describes the intensity and color information for each pixel on the monitor. There will rarely be a one-to-one correspondence between the raw image pixels and the monitor pixels so the display array will have to be interpolated from the original array. Alternatively, if the image is to be printed on a solid medium such as paper or film, then the array of pixel information is converted into a new array that describes the intensity and color information for each printing element. On a sophisticated inkjet printer there may be ten different inks available and the print head may be capable of ejecting hundreds of separate ink droplets per centimeter of print medium. The data array that is required for printing is thus very much larger than the original image array. It contains a lot of information very specific to the particular image output device, but it need only exist for the duration of the printing process and need not be stored long term.

## 2.3   Image Information

It should now be quite clear that because digital images are so easily stored, transmitted, and displayed on different media the physical form of a specific digital
image is highly context-dependent. Much more significant than the physical form
of a digital image is its *information content*. The *maximum* amount of information that can be stored in an image depends on the number of pixels it contains
and the number of possible different intensities or colors that each pixel can have.
The *actual* information content of the image is invariably less than the maximum
possible. As well as the uncertainty in the spatial origin of the signal due to the
point spread function, there will be some uncertainty about the reliability of the
intensity or color information due to a certain amount of *noise* in the measured signal.

When we perform image processing we are sorting and manipulating the information in an image. Often we are trying to separate certain parts of the 'true' signal
from the noise. In doing this we must be careful not to accidentally destroy important information about the imaged subject, and also not to introduce new noise or
artifacts that might be accidentally interpreted as information.

### 2.3.1   Pixels

You might say that the fundamental particle of digital imaging is the pixel – the
smallest piece of discrete data in a digital image. The pixel represents discrete *data*,
not necessarily discrete *information*. Due to the point spread function, subject movement, and several other effects, information from the imaged object will to some
extent be distributed amongst adjacent pixels (or voxels). When discussing color
images we could separate the individual color components of each pixel (e.g. the
red, green, and blue data that describe a pixel in an RGB image) but since we are
mainly dealing with gray scale images in medical imaging we need not worry about
this refinement here. However, we do have to be careful about the way we use the
term 'pixel' in digital imaging, even after defining it as a 'picture element'. *Pixel*
can mean different things in different contexts and sometimes conflicting contexts
are present simultaneously.

A pixel might be variously thought of as:

1. A single physical element of a sensor array. For example, the photosites on a
   semiconductor X-ray detector array or a digital camera sensor.
2. An element in an image matrix inside a computer. For an $m \times n$ gray scale image
   there will be one $m \times n$ matrix. For an $m \times n$ RGB color image there will be three
   $m \times n$ matrices, or one $m \times n \times 3$ matrix.
3. An element in the display on a monitor or data projector. As for the digital color
   sensor, each pixel of a color monitor display will comprise red, green and blue
   elements. There is rarely a one-to-one correspondence between the pixels in a

digital image and the pixels in the monitor that displays the image. The image data is rescaled by the computer's graphics card to display the image at a size and resolution that suits the viewer and the monitor hardware.

In this book we will try to be specific about what picture element we are referring to and only use the term pixel when there is minimal chance of confusion.

### 2.3.2  Image Size, Scale, and Resolution

Shrinking or enlarging a displayed image is a trivial process for a computer, and the ease of changing the displayed or stored size of images is one of the many advantages of digital imaging over older film and paper based technology. However, technology changes faster than language with the result that terminology, such as references to the *size*, *scale* and *resolution* of an image, can become confused. We may not be able to completely eliminate such confusion, but being aware of the possibility of it should make us communicate more carefully. We may need to be explicit when we refer to these characteristics of an image, and we may need to seek clarification when we encounter images which are described with potentially ambiguous terms.

What is the *size* of a digital image? Is it the image matrix dimensions, the size of the file used to store the image, or the size of the displayed or printed image? The most common usage defines *image size* as the rectangular pixel dimensions of the 2D image – for example $512 \times 512$ might describe a single slice CT image. For very large dimension images, such as digital camera images, it is common to describe the image size as the total number of pixels – 12 megapixels for example.

*Image scale* is less well-defined than image size. In medical imaging we generally define the *Field of View* (FOV) and the image matrix size. Together these define the *spatial resolution* of the raw image data. We discuss spatial resolution in detail in Chapter 3. Many file storage formats include a DPI (dots per inch) specification which is a somewhat arbitrary description of the *intended* display or print size of the image. Most software ignores the DPI specification when generating the screen display of an image, but may use it when printing.

### 2.3.3  Pixel Information

If, as in most cases, an image represents the state of a subject at some time in the past (e.g. A photograph, a CT scan, an MR image), then the image data represents a discrete sampling of some physical property of the subject. An MR image, for example, will have been acquired with a specific field of view and matrix size. A pixel in the raw MR image data represents the average MR signal intensity in a specific volume of space inside the MR scanner (together with a certain small amount of neighboring pixel information according to the point spread function).

The precision with which the signal intensity is measured and recorded, and the amount of noise, determine the maximum possible information content of the image data.

### 2.3.3.1  Bit Depth

An image must have adequate spatial resolution to show the spatial separation of important separate objects. It must also have adequate *intensity* resolution, or precision, to record any contrast difference between objects – assuming there is a measurable difference in the signals from the objects. In a measured image individual pixels represent discrete samples of the spatially continuous measurement signal. The digital encoding of the measured signal intensity for each pixel also has discrete rather than continuous values – the measured signal is *quantized*. The number of discrete levels, the maximum precision of the stored intensity data, is defined by the number of bits used to store the data – the *bit depth*. The *actual* precision of the data will be limited by the measurement hardware and system noise.

The choice of bit depth used to store raw image data is generally based on the precision of the measurement system. In a properly engineered imaging system we want the precision of the data recording system to be a little bit greater than the precision of the physical measurent apparatus. If the recording precision were too low then expensive measurement hardware would be inadequately utilized and potentially useful information would be lost in the data recording process. Alternatively, if the recording precision were excessively high, no extra information would be saved but data storage space would be wasted, and both data transmission and image processing would be slower.

By way of example, consider the data precision requirements of CT. In CT images each pixel stores a calculated integer CT number which can range from $+3,000$ for dense bone to $-1,000$ for air. We thus need a bit depth that will encode at least 4001 CT numbers. The bit depth required is 12 ($2^{12} = 4,096$). Typical bit depths for other imaging modalities are 10 or 12.

All imaging data is measured and stored with much higher precision than a human can actually see. Human visual perception has quite poor and non-linear discrimination of light intensity. By some estimates humans can reliably distinguish only about 32 different gray scale levels. This is clearly demonstrated in Fig. 2.3 where we see that even if we reduce the number of distinct gray scale levels from 256 to 16 the effect is barely noticeable. Most gray scale image display devices, for example monitors, have a bit depth of eight, with the result that $2^8 = 256$ different intensity levels can be displayed. There are two apparent paradoxes here. Firstly we acquire data with a precision of $2^8$ or higher, secondly we display this data with a precision of $2^8$, and yet we can only see with precision $2^5$. Why do we bother to record and display images with such an apparent excess of intensity precision?

Remember that the raw data we acquire represents the variation in intensity of some measurable physical phenomenon. The information of interest, perhaps some anatomical details, will probably not be represented by intensity variations across

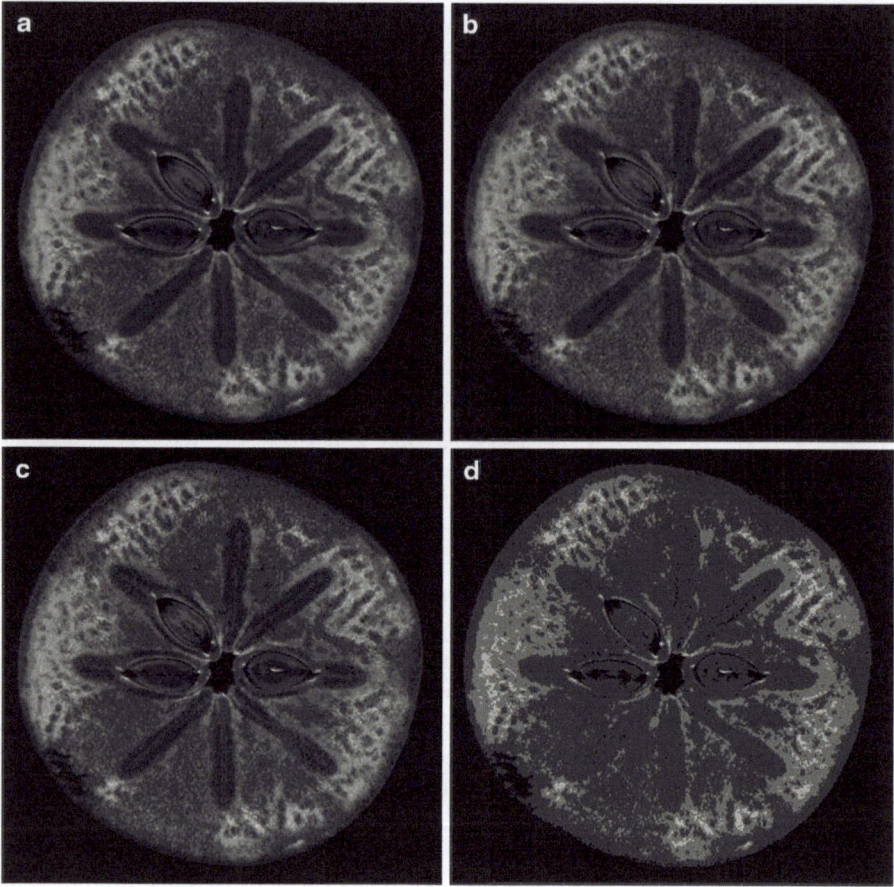

**Fig. 2.3** The information content of a digital image depends on the number of pixels and the number of distinct intensities. This figure illustrates the effect of reducing the number of intensity levels on image information content. These MR images of an intact persimmon have 256, 16, 8, and 4 distinct gray scale intensities (**a–d** respectively). In this particular image the reduction in displayed intensity precision from 256 to 16 is barely noticeable. This may not be the case in all images, and in some cases important information could be lost by such a reduction in precision – particularly if we want to see the details in a small region or identify very subtle changes

the full range of measured intensities. It is often impossible or impractical to predict the intensity range of interest prior to acquisition. Thus the imaging technology must be able to measure a range of intensities that can be reliably predicted to include the information of interest, *and* it must record this range with sufficient precision (i.e. intensity resolution) to enable post-acquisition *expansion* of this range to create a display for human vision. That display must have sufficient intensity contrast detail to enable reliable interpretation

Because it is often difficult to predict the intensity range of the information of interest within the raw intensity data one of the most common image processing adjustments is the selective and interactive improvement of contrast performed while viewing an image. The raw information encoded in small differences of intensity may be imperceptible to the human viewer until these differences are exaggerated by contrast enhancement.

In CT data we usually can predict the range of CT numbers that will cover the information of interest for a particular investigation. In this case it is normal practice to use a standard *Window Function* to select a specific range of CT numbers to be displayed as an 8 bit gray scale image as illustrated in Figs. 2.4 and 2.5.

It is important to remember that no amount of post-acquisition contrast enhancement will be able to extract a difference that is not present and significant in the recorded physical phenomenon. As we shall see, there are a number of image processing 'tricks' we can perform to increase the *apparent* differences, and there are even some 'built in' to the human vision system. Whether such information is present and significant depends on the precision and noise level of the image acquisition and recording system.

Although we demonstrated in Fig. 2.3 that reduction of *displayed* intensity precision may be imperceptible it does not follow that we can reduce raw data precision with impunity. When we apply contrast enhancement to improve the visibility of displayed contrast the desired information must be available in the precision of the raw data.

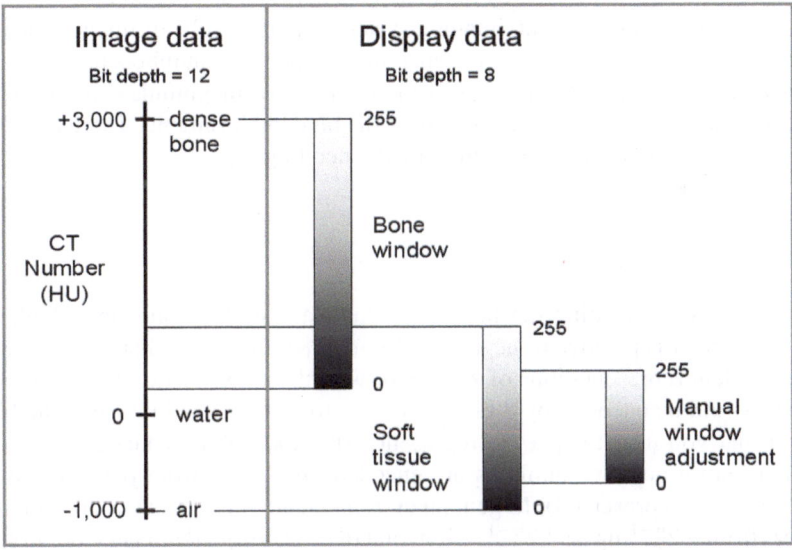

**Fig. 2.4** Human perception cannot resolve the full precision of stored CT image data (typically 12 bits). According to the anatomy of interest, standard *Window Functions* are used to select a defined range of data for display. The precision of the display data is 8 bits. A subset of this data may be selected manually by the viewer to further enhance the visibility of specific anatomical features

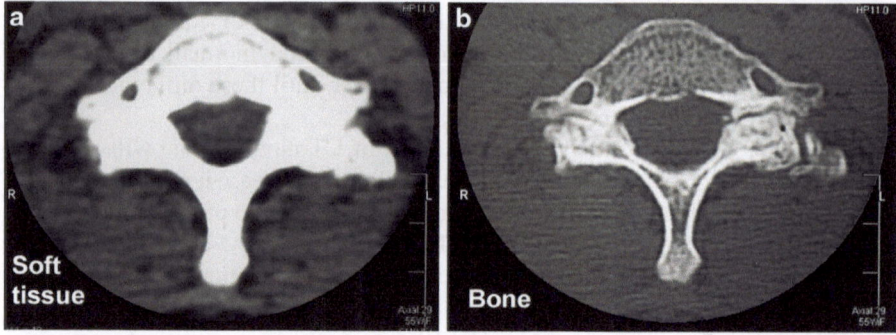

**Fig. 2.5** Specific ranges (windows) of CT image data are used to display maximum contrast according the anatomy of interest. Here a single raw data set has been windowed for soft tissue (**a**) and bone (**b**)

## 2.3.4  Ways of Representing Numbers

So far we have described the binary representation of image data only in terms of positive integers. Using 8 bits we can represent (encode) all the integers from 0 to 255, with 14 bits all the integers from 0 to 16383, and so on. This is fine for image data that is naturally described by positive integers, such as pixel intensities, but often the raw data acquired by an imaging system and the results of image processing are *not* simple positive integers. They may include negative numbers (e.g. voltages), decimal fractions, and may range over many orders of magnitude – more than we can represent using positive integers within the bit depth available. There are several ways of addressing these needs using binary encoding.

### 2.3.4.1  Signed and Unsigned Integers

In *signed integer* encoding the first bit of the available bits indicates whether the encoded number is positive or negative. You might at first think that this will lead to two equivalent representations of zero ($\pm 0$) with the result that only $2^8 - 1 = 255$ numbers could be encoded by 8 bits. However, for 8 bit signed integers, the binary number that you might expect to represent $-0$ (1000 0000) in fact encodes $-128$ (this is because negative numbers are encoded differently from positive numbers and '$-0$' is not represented). In general an $n$ bit signed integer can represent all the integers from $-2^{n-1}$ up to $+2^{n-1} - 1$, a total of $2^n$. *Unsigned integers*, the first type of binary encoding we discussed, can represent all the integers from 0 up to $2^n - 1$ with $n$ bits.

### 2.3.4.2 Floating Point

In floating point encoding numbers are represented by a binary equivalent of the decimal 'scientific notation'. For example, the decimal scientific notation for the number 123456 would be $1.23456 \times 10^5$. In floating point encoding this is changed to $0.123456 \times 10^6$. The *significand* (0.123456) and the *exponent* (6) are stored side by side as signed integers. Notice that the significand is actually *not* an integer – the decimal value of the binary number is always interpreted as a number between 1.0 and 0.1. There are many different floating point conventions that assign different bit depths to the significand and the exponent according to the need for precision (bit depth of significand) or dynamic range (bit depth of exponent).

## 2.3.5 Data Accuracy

What about data *accuracy*? Storing data in a large file with a high bit depth does not mean the recorded measurements are accurate. Nor does it *guarantee* that they are precise. A noisy or unstable imaging system will not be precise, and an uncalibrated system will not be accurate. Precision and accuracy are two distinct properties of measurement. Have a look at Appendix B if you are unsure of the difference.

By data accuracy we mean how well does recorded intensity information, whether relative contrast or an absolute measurement with specific units, reflect the actual physical properties of the imaged object. We would also like the intensity information to be spatially reliable, in other words, it can be attributed to a well-defined region of space.

Spatial accuracy is not strictly a property of individual pixels in the image data. A loss of spatial accuracy means that the image data attributed to a specific region of space in fact contains some contributions from adjacent regions. This could be a result of the Point Spread Function mentioned previously, or movement of the imaged object during the period of measurement. The measurement of the blurring aspect of spatial inaccuracy is discussed in terms of the Modulation Transfer Function (MTF) in Chapter 3.

In CT the calculated and stored CT numbers are directly related to the linear attenuation coefficient ($\mu$) of the imaged tissue. A CT system needs regular calibration using a phantom containing regions of well defined attenuation coefficient to ensure that the calculated CT values are accurate. In other modalities, e.g. MRI and plain X-ray radiography, we are usually measuring relative intensities of signals rather than absolute physical properties. Such systems still require calibration to check the spatial accuracy of the data.

## 2.4   Image Metadata

If the only data we stored in a digital image file was a long sequence of bits representing pixel intensities we would be missing a lot of essential information about the image. We would not even be able to display the image if we did not have a record of the pixel dimensions $m$ and $n$. We would not know if the data represented a gray scale or color image. Other important information such as who or what was imaged, and how and when the imaging was performed would also have to be recorded somewhere and reliably connected with the pixel intensity data. It makes sense to store this sort of information, and a lot more, together with the pixel intensity data in a single image file. With a few rare exceptions this is the basic format of all digital image files. All the non-intensity data is called the *image metadata*, or image file *header*.

### 2.4.1   Metadata Content

An image file header is not necessarily a sequence of bytes with conventional text encoding. The header itself has a structure that is specific to the file type. How then does a piece of software know what kind of image file it is trying to read? Usually the first two bytes of the header itself are a 'signature' that defines the image file type. You can see this by opening an image file with a basic text file editor. Most of the displayed symbols will be meaningless because it is not text code, but the first few characters include text that indicates the file type (only try this with a very small image file or the text editing software may fall over).

The contents and format of the metadata depend on the particular image file type but *always* includes essential information including the size of the image matrix ($m$ and $n$) and the precision (bit depth). An image file from a digital camera will usually include metadata that describes the camera settings for that particular image (Fig. 2.6). You can inspect some of an image file's metadata without displaying the image (use File Properties in Microsoft Windows). Any image display software *must* read some of the metadata before it can work out how to display an image.

If the data is compressed then the metadata needs to describe the compression method and the parameters used. A medical image file will include information about the scanner on which the image was acquired, the acquisition parameters, and a way of identifying the patient. For privacy and efficiency, personal and clinical data are usually stored in a file separate from the image file. Figure 2.7 gives a schematic representation of the separate components of a simple digital image file. An example of some typical header information from a medical DICOM format image is shown in Fig. 2.8.

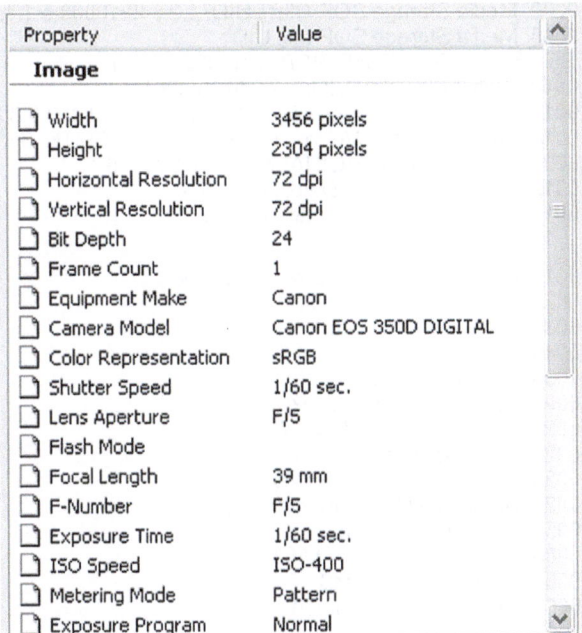

| Property | Value |
|---|---|
| **Image** | |
| Width | 3456 pixels |
| Height | 2304 pixels |
| Horizontal Resolution | 72 dpi |
| Vertical Resolution | 72 dpi |
| Bit Depth | 24 |
| Frame Count | 1 |
| Equipment Make | Canon |
| Camera Model | Canon EOS 350D DIGITAL |
| Color Representation | sRGB |
| Shutter Speed | 1/60 sec. |
| Lens Aperture | F/5 |
| Flash Mode | |
| Focal Length | 39 mm |
| F-Number | F/5 |
| Exposure Time | 1/60 sec. |
| ISO Speed | ISO-400 |
| Metering Mode | Pattern |
| Exposure Program | Normal |

**Fig. 2.6** Image *metadata* is associated (and usually stored with) pixel intensity data. The metadata describes how to display the pixel data, and may include information about the method of data acquisition and the image subject. This table shows metadata retrieved from a digital camera image file by the Microsoft Windows *File Properties* command

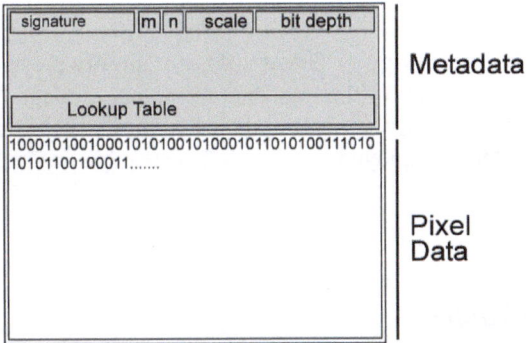

**Fig. 2.7** Schematic representation of a digital image file. The metadata describes the image geometry, the source and acquisition parameters, details of compression if any, and may include a lookup table or color map describing the display intensity/color for the stored data. The data section contains the actual, often compressed and encoded, information about the pixel intensities and color

```
0002,0002  Media Storage SOP Class UID: 1.2.840.10008.5.1.4.1.1.4
0002,0003  Media Storage SOP Inst UID:
2.16.756.5.5.100.1702245055.21375.1214885758.2.1
0002,0010  Transfer Syntax UID: 1.2.840.10008.1.2.1
0002,0012  Implementation Class UID: 1.2.276.0.7230010.3.0.3.5.3
0002,0013  Implementation Version Name: OFFIS_DCMTK_353
0008,0008  Image Type: ORIGINAL\PRIMARY\OTHER
0008,0012  Instance Creation Date: 20080704
0008.0013  Instance Creation Time: 095526
[.....]
0010,0010  Patient's Name: RatBrain01072008
0010,0020  Patient ID: RatBrain01
0010,0030  Patient's Birth Date:
0010,0040  Patient's Sex: F
0010,1030  Patient's Weight: 5
0018,0020  Scanning Sequence: RM
0018,0021  Sequence Variant: NONE
0018,0022  Scan Options:
0018,0023  MR Acquisition Type: 2D
0018,0024  Sequence Name: m_msme (pvm)
0018,0050  Slice Thickness: 0.5
0018,0080  Repetition Time: 500.553
0018,0081  Echo Time: 19.0746
0018,0083  Number of Averages: 1
[.....
```

**Fig. 2.8** Part of the metadata (header) of a DICOM format image file. This particular file is from a magnetic resonance microimaging system. This part of the header describes the file type, when and where the image was acquired, the acquisition parameters, and some details describing the sample or patient. The eight digit numbers on the left are standard DICOM labels for general and modality-specific image information

In ImageJ use Menu: Image > Show Info... or simply press 'I' on the keyboard to show some or all of the metadata for an open image file The part of the metadata shown by this command depends on the file type. For DICOM files the full header is displayed.

## 2.4.2 Lookup Tables

If we measure a signal with 12 bit precision then the most obvious way to store the data would be as a list of pixel intensities, each using 12 bits of storage space. While this is the normal way to store raw image data it is often not the most efficient. Even in a high precision data set it is likely that there are far fewer measured intensities than possible intensities. Consider an 8 bit gray scale image that contains only 31 different measured signal intensities. We may still need 8 bit precision to accurately describe the *relative* differences between the intensities, but we actually only have to *store* 31 different intensities values. A good way to save storage space for such an image is to include a *Lookup Table* (or LUT) in the file header. In our example image the lookup table would list the 31 different intensites with 8 bit precision and each would have an associated 5 bit *index*. Instead of storing all the pixel intensities with the full 8 bit precision we could store a 5 bit index for each image pixel ($2^5 = 32$). This method would require only $\frac{5}{8}$ of the intensity data storage space.

Lookup tables are also referred to as *Color Maps* or *Color Palettes*. Color image files that use lookup tables are called *Indexed Color* images. Image files that do not use a lookup table and store individual pixel data with full precision are called *True Color* images.

Because the lookup table is distinct from the pixel intensity data the way image data is displayed can be easily and conveniently changed by manipulation of the lookup table without having to adjust the individual pixel intensity or color data. A color lookup table can also be used to display gray scale image data as a *'false color'* image (Fig. 2.9).

**ImageJ.** Use Menu: Image > Lookup Tables to change or invert the lookup table for an image.

Lookup tables are also used to adjust the output of display hardware. A typical computer graphics card (display adapter) includes a built-in lookup table that adjusts the raw display data to suit the specific monitor attached to the card. Monitor calibration systems adjust these lookup tables in order to produce a defined monitor light output (color and brightness) as measured by a photometer placed on the monitor face.

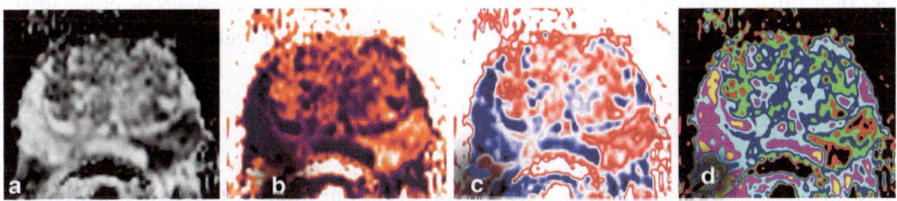

**Fig. 2.9** A Lookup Table may be part of the image file metadata and specifies how to *display* the raw image data. In this example a (8 bit gray scale) diffusion weighted MR image of a human prostate (**a**) is displayed using three different color lookup tables. Creation of similar 'false color' images can sometimes increase the visibility of subtle diagnostic features present in medical images (see Fig. 6.8 for a graphical display of the 'Union Jack' lookup table data used for image **c**)

## 2.5 Image Storage

The storage and transmission of medical images is obviously of critical importance to medicine. Images must be stored safely to protect both the integrity of the data and the privacy of patients, but images also need to be easily available when and where they are needed by medical staff. The imaging software user will normally have the option to store processed images in a number of different standard formats. Although the methods of transmission of images are generally not of concern to image acquisition and processing it is important to be aware that the time required for transmission depends on the size of the image file.

### 2.5.1  Image File Formats

The choice of image file format has implications for:

1. The size of the stored image file
2. The type and amount of metadata that can be stored
3. The availability of multiple layers and transparent layers
4. The flexibility or 'customizability' of the content
5. The integrity of the data
6. The speed of image transmission
7. Software compatibility

All general-purpose file formats are designed to handle color images. The medicine-specific formats, e.g. DICOM, are primarily designed for gray scale images but are flexible enough to store color images when necessary. The following is a simplified overview of some common image file formats. Most of these formats utilize or enable image data compression.

Image data compression methods are categorized as either *lossless* (no intensity or color information is lost in compression), or *lossy* (some intensity or color information is lost). Section 2.5.2 below discusses compression methods in more detail.

#### 2.5.1.1  Bitmaps and BMP Files

The simplest and most obvious way to store a digital image of size $m \times n$ pixels is as an $m \times n$ array of pixel intensities – commonly referred to as a *bitmap*. You can think of a bitmap as a table in which each entry represents the intensity of a pixel. For a gray scale image with 256 possible intensities we will need $m \times n \times 8$ bits to store the image data.

The term 'bitmap' has both a generic and a specific common usage. The generic term refers to all digital images that are represented as spatial maps of pixel intensities, in other words, as arrays in which each array element corresponds to a discrete position in space. The term *raster graphics* is also used for these images. The specific usage of 'bitmap' refers to a particular file format – *Windows Bitmap or BMP*.

### 2.5.1.2 Vector Graphics

An alternative method of encoding some types of digital images is *vector graphics*. A vector graphics image describes the line and tonal detail as a collection of vectors – lists of points that describe the geometry of objects in an image. Only when the image is displayed or printed is a raster graphics (bitmap) image generated from the vector graphics information – the vector data is *rasterized*. This method is efficient for storage of synthetic images created with graphic design tools as it provides a precise and easily scaleable description of geometrical image features. It is also good for animations as the changing composition of the image (objects, perspective, shadows, etc.) can be calculated geometrically from the virtual objects.

Vector graphics is unsuitable for representation of images with subtle tonal detail, such as anatomical medical images, but would be suitable for the masks and line diagrams used in medical treatment planning. In contrast to rasterization, the reverse process of converting a bitmap or raster graphics image to vector graphics (vectorization) is a relatively very difficult process that is likely to lead to significant loss of image information.

### 2.5.1.3 JFIF (JPG)

What we commonly call JPG or JPEG images (with file names ending in .JPG) are really JFIF (*JPEG File Interchange Format*) files. JPEG is a compression method, not a file format, and it may be used within file formats other than JFIF, such as TIFF. The JPEG algorithm (outlined below) provides efficient and controllable compression of images but it is most often implemented via a *lossy* method, meaning image information is discarded in the compression process. Any lossy compression process needs to be used with extreme caution on medical images in case important clinical information is lost.

### 2.5.1.4 GIF

The *Graphic Interchange Format* (GIF) is ideal for storage of simple images containing few distinct colors and very limited tonal detail. Only 256 different colors may be stored and these are encoded in a lookup table. GIF provides for multiple layers, including transparent layers. Transparency permits an image to be displayed on a background such that pixels designated as transparent in the image are displayed with the background color. The background could be a solid color or another image. The layers in a GIF image can be displayed in a timed sequence enabling simple animation. The very limited intensity precision of GIF makes it unsuitable for anatomical medical images.

### 2.5.1.5  PNG

The *Portable Network Graphics* (PNG) file format was developed as a lossless storage format that would still provide efficient compression. PNG provides for variable precision (8–16 bits) and variable transparency, but does not allow multiple layers. Medical examples of the use of variable transparency would be the overlaying of a color treatment plan on an anatomical image, and the superposition of two images of the same subject acquired from different imaging modalities – CT and PET, say. Because of its high precision and lossless compression PNG could safely be used for storage and transmission of individual medical images. The PNG file will, however, lack the extensive and standardized metadata capability of the DICOM format.

### 2.5.1.6  TIF

The *Tagged Image File Format* (TIFF or TIF) was designed by developers of color printers, monitors, and scanners. It focuses on the quality of the image rather than the size of the image file, however several different compression methods are supported. A useful feature of the TIF format is that it can store multiple images, or layers, in a single file. Such multiple layers might, for example, represent images of the same object acquired at different times or with different techniques, or an anatomical image and a separate set of annotations. In digital cameras the TIF format is commonly used to store uncompressed image data together with a small JPEG-compressed 'thumbnail' image bundled together in a single file (this is the EXIF file structure). The thumbnail image allows a preview of the main image without the need for decompression of the full image data.

The TIF format can be thought of as a package for one image or a collection of images. Depending on the software a range of compression methods, both lossless and lossy, may be available when saving an image in TIF format. A disadvantage of the flexibility of the TIF format means that TIF files created with one type of software may not be readable by some other software. This is the usual reason for the 'Unsupported Tag' error message which sometimes appears when unsuccessfully trying to open a TIF file.

### 2.5.1.7  DICOM

Most medical imaging systems archive and transmit image data in DICOM (*Digital Imaging and Communications in Medicine*) format. The DICOM standard (www.nema.org) is designed to enable efficient exchange of radiological information (images, patient information, scheduling information, treatment planning, etc.) independent of modality and device manufacturer. When we talk about a 'DICOM image' we mean an image file that conforms to Part 10 of the DICOM standard, which currently has 18 parts.

A DICOM image file comprises a header (Fig. 2.8) of image metadata and the raw image data within a single file. The header contains information about the imaging system, the acquisition parameters, and some information about the patient (or the object that was imaged). The DICOM standard provides for lossless and lossy JPEG compression, and other lossless compression formats. Multiple frames, such as the contiguous slice images of a 3D data set, can be stored in a single DICOM file. An important feature of the DICOM format is its ability to store pixel intensity data with precision of 8, 12, 16, or 32 bits according to the measurement precision of the imaging system.

A collection of DICOM files representing all the images acquired from a patient in a single examination usually includes a separate DICOMDIR file that acts as a stand alone 'superheader' describing the individual DICOM image files which have unhelpful file names that make no sense without the DICOMDIR file. Sometimes you can inspect the header information of a single DICOM file to work out what the image represents.

## 2.5.2 Image Data Compression Methods

Many image file storage formats compress the image data to reduce storage space requirements and speed image transmission. Most image processing software permits the user to specify whether or not to compress the data and what compression method to use. As mentioned above, the choice of method may be based on software compatibility but also on whether any loss of information can be tolerated. Lossy data compression methods discard the information that is considered to be least obvious to human perception and can often achieve an 80–90% reduction in file size. Lossless compression methods reduce the size of the stored data by methods that are perfectly reversible – they eliminate *only* redundant data. Images stored with lossless compression methods are identical in information content to their uncompressed counterparts.

Three different kinds of redundancy are possible in image data:

1. Coding redundancy. This is the type of redundancy described above where the data encoding method has more precision than is necessary for a particular image. This type of redundancy can be addressed by using a reduced bit depth and a lookup table.
2. Spatial redundancy. This occurs when there are large regions of identical pixels each containing identical information – for example the black background of an X-ray image. This redundancy can be reduced by a method that encodes the description of homogeneous regions.

3. Information redundancy. This is information that cannot be perceived – for example spatially small regions with very small differences in intensity and color cannot be seen by humans. This redundancy can be eliminated by making such regions homogeneous. Note that such newly homogenous regions will then be amenable to elimination of spatial and coding redundancy.

A bitmap file format stores an $m \times n$ image as an $m \times n$ rectangular array of pixel intensities. There are two reasons why this format is generally a very space-expensive way to store an image. Firstly, most images have significant areas of indentical, or nearly identical, pixel values. This is spatial redundancy. In a medical image, for example, most of the background is usually black, or contains only noise. The second inefficiency lies in the fact that there are often far fewer different pixel intensities present in the image than can be encoded with the nominal bit depth – there is more precision available than is necessary to encode the actual information in the image. This is coding redundancy.

We can drastically reduce the amount of media space required for image storage (and reduce the time required for image transmission) by reducing the redundancies just mentioned. If the first 100 rows of an $m \times n$ image matrix all represent black background then instead of using $100 \times n \times 8$ bits, all set to zero, to store this information we could simply use a code that says 'pixels 1 to $100n$ have value zero'. Not only would this encoding save a huge amount of space but it results in no loss of image information. Alternatively, if we are prepared to lose some information considered to be unimportant, then we might decide to adjust very similar pixel values to make them identical and thus reduce the total number of different intensities we need to encode. When the image contains *fewer* discrete intensity values than the nominal bit depth can encode we can save space by encoding the intensities in a lookup table.

Image data compression methods take advantage of the spatial, intensity, and information redundancy just described. Statistical analysis of the image data can lead to further improvements in compression. If we make the assumption that the least common pixel intensities do not represent significant image information then we can omit them from the lookup table by changing them to the closest more common value. Similarly, we might decide that single pixels, or small groups of pixels, that do not fit some measured pattern or trend found in their neighborhood are not important and replace their original values in the stored encoding. The more assumptions of this kind we make the more space we save, but more original image information is lost.

### 2.5.2.1   JPEG

The JPEG (Joint Photographic Experts Group) compression method is ubiquitous in digital imaging. In fact it is so common that the name of the method is used more commonly than the name of the main file format (JFIF) that uses the JPEG compression method.

The JPEG compression algorithm includes both lossless and lossy steps. The lossy step exploits the limitations of human vision and reduces the precision of that part of the image information which is most weakly perceived by the eye. Specifically, this is small differences in intensities between closely spaced pixels (in technical terms: reduced precision of high spatial frequency components. We will have a lot more to say about spatial frequency in Chapter 4). The method breaks images down into blocks of $8 \times 8$ pixels and reduces the information content of each block. Because the blocks are processed independently, obvious discontinuities appear at the block edges in highly compressed images (Fig. 2.10). The appearance of the characteristic square pattern should not be confused with *pixelation* which results from simple duplication of pixels in images enlarged by the nearest neighbor method (Chapter 8).

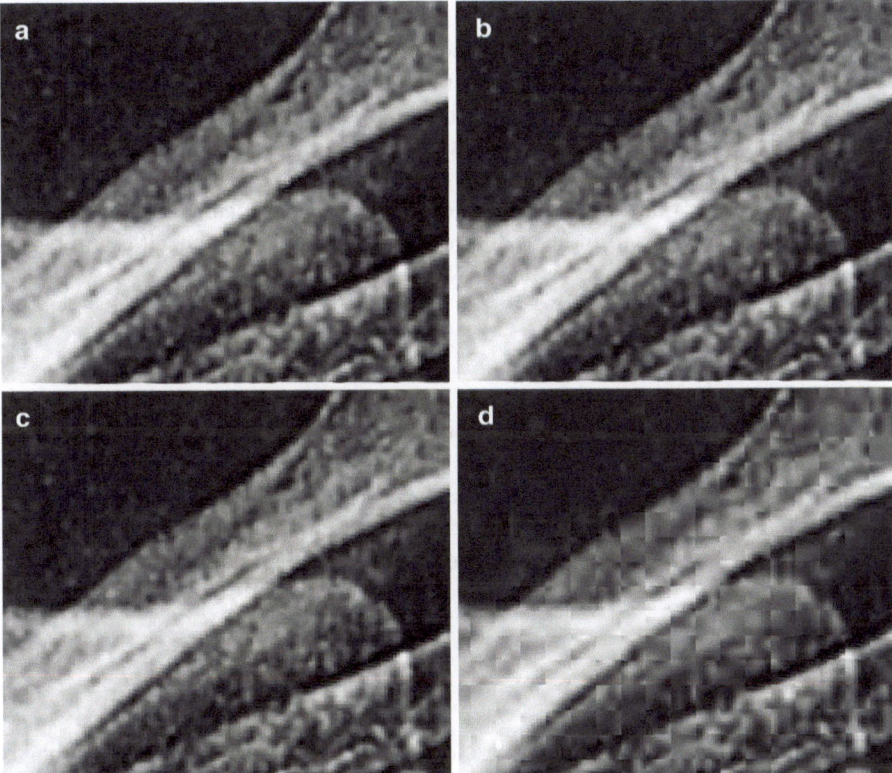

**Fig. 2.10** Plain X-ray image illustrating the effect of different levels of JPEG compression. (**a**) Original image. (**b**) JPEG compression level 12 (minimum compression). (**c**) JPEG compression level 6 (medium compression). (**d**) JPEG compression level 0 (maximum compression). At high levels of compression the independently processed $8 \times 8$ pixel regions become distinctly visible and edge features are severely degraded. In less severely compressed images a more subtle speckled 'halo' artifact may be visible along edges

In the context of medical imaging we would not usually store raw image data using the lossy JPEG compression method. We might, however, choose to save a *copy* of an image in lossy JFIF format in order to send it as a reasonable-sized file over the internet. In this case both the sender and the recipient need to be conscious of the possibility that the JFIF version of the image may lack some information that is present in the raw data. 'Lossless' JPEG compression is available in some software.

A modified method, JPEG2000, has been developed to address some of the deficiencies of the original JPEG method. It is not yet in common usage and cannot be decoded by most image display and processing software, however, it is supported by the DICOM standard.

### 2.5.2.2  Packbits

Packbits is a lossless compression method that uses *run length encoding* (RLE). RLE takes advantage of the fact that most images have long lines of adjacent pixels of identical intensity or color. Rows of an image matrix are broken up into packets each with its own 'mini-header'. Whenever three or more adjacent pixels are identical the mini-header is set to indicate that the packet describes how many pixels of a single specific color and intensity follow. If adjacent pixels are not identical the packet header is set to indicate that the packet describes a specific number of non-identical pixels.

### 2.5.2.3  ZIP, PKZIP

ZIP (derived from PKZIP) is a general purpose lossless data compression method. It can be used within an image file format (e.g. TIFF), as a method to compress a single file, or to convert a series of files and directories (folders) into a single compressed archive file. Most common software supports the 'unzipping' of ZIP compressed files but not all image processing software can open image files with internal ZIP compression.

### 2.5.2.4  LZW

LZW (Lempel-Ziv-Welch) is a lossless compression method used in GIF images and is available in some TIFF implementations. It reduces spatial redundancy by creating a 'dictionary' of common intensity or color patterns (a sophisticated kind of lookup table), and then encoding the image data as a sequence of dictionary references.

### 2.5.2.5 Which Method Is Best?

We have looked at some of the most common file formats and image data compression methods. Figure 2.11 compares the file sizes and image quality for several of these. There are a few important points to notice: (1) In the absence of compression there is little differences in file sizes. (2) Both the type of image data and the compression method have a major effect on the compressibility. (3) The lossy JPEG method does not always provide greater compression than lossless methods. (4) Lossy methods do not *always* produce visible degradation of image quality.

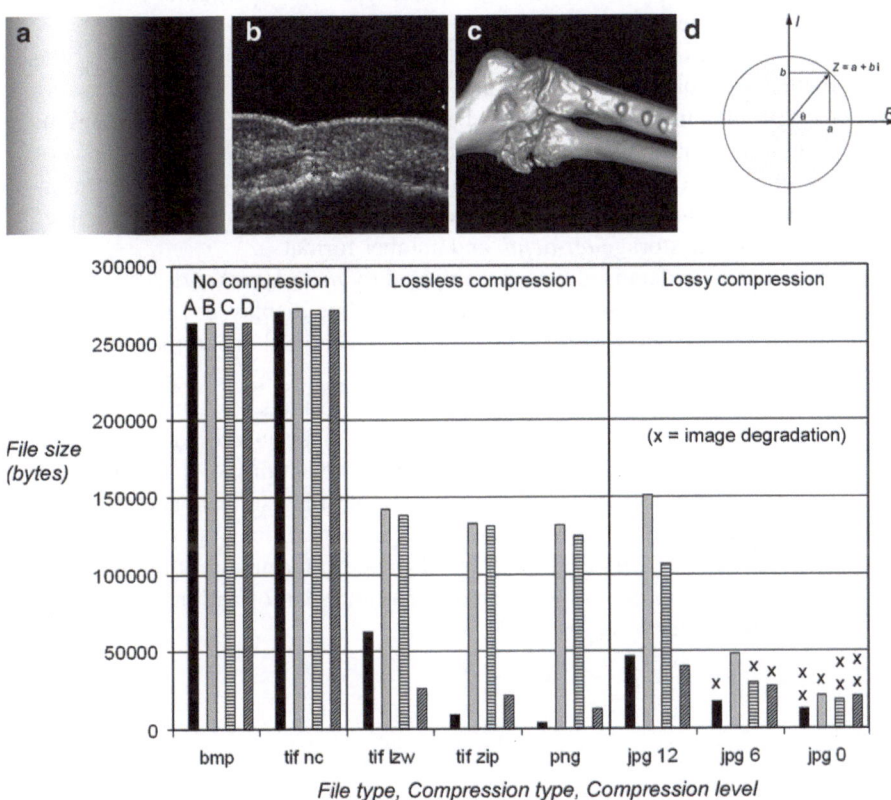

**Fig. 2.11** Comparison of file sizes for different image types, file types, and compression methods. The 'compressibility' of an image depends on its content and the compression method. In the case of lossy compression the severity of image degradation depends on both the degree of compression and the image content. Points to note: Lossless PNG compression outperforms even the lossiest JPEG compression for images (**a**) and (**d**). The ultrasound image (**b**) is only slightly degraded even at the highest level of JPEG compression (All original images were 512 × 512 pixels, bit depth 8, gray scale. The compressed images are not shown.)

## 2.6 Summary

- A digital image is an *encoding* of an image amenable to electronic storage, processing, and compression. A digital image typically represents a discrete and regular rectangular sampling of some property of physical space.
- A *pixel* (picture element) can be either: an element in the raw image data; an element in an image data array created from raw data; or the smallest element in a display device such as a monitor. There may not be a one-to-one correspondence between these pixel types for a particular image at a particular time.
- The *information content* of a digital image is inherently limited by the method of acquisition and storage of the raw image data. Image processing can filter the information content but cannot increase it. The maximum information content that can be stored is limited by the pixel dimensions of the image and the intensity resolution or bit depth. Signal intensity information is *quantized* in digital image data. It can only have discrete values limited by the storage *bit-depth* and number format.
- Pixel intensity data is always accompaied by *metadata* that describes how the intensity data is stored and how it should be displayed. Metadata may also include details of the method of image acquisition and the subject of the image.
- Digital images can be stored in numerous formats, and interconverted between formats, according to the relative importance of data integrity, file size, speed of transmission, and the conventions of particular image processing software. Many digital image files store image data in a compressed format for space and transmission efficiency. Compression methods may be *lossless* (all original image information is retained) or, to a variable degree, *lossy* (some original image information is discarded to enable greater compression).

# Chapter 3
# Medical Images

## 3.1 Introduction

Most of the attributes of digital images and the methods of image processing introduced in this text originate from outside of medicine. We can single out medical imaging for special consideration because the lives of people often depend on correct acquisition, processing, and interpretation of medical images. It is important that individuals responsible for acquiring and processing medical images understand both the nature of the raw material they work with, and the way the images they produce will be used. Those using medical images for research, rather than purely clinical purposes, also need to understand the way their raw data is acquired to ensure the scientific rigor of their work.

The purpose of medical imaging is to reveal and record the *structural* or *functional* state of the body. Mostly we want to see what is going on inside the body – to check that all is well, or to find out why all is not well. Sometimes we want a record of the current state of the body to be referred to at some future time – to monitor the progress, or absence of progress, of a disease or a treatment.

The majority of medical images are intended to reveal aspects of the body that cannot be observed by visual inspection or physical examination of the exterior of the body. Interestingly, most medical imaging methods produce a visible light image that *represents* a physical property of body tissue which *cannot* be observed with visible light – at least not without resorting to surgery, which is mostly expensive and possibly counterproductive. We know from years of development of medical imaging modalities that certain measured physical phenomena correlate with biological properties of interest – either disease or normal structure and function.

The correlation between physics and biology is easy to understand in the case of something like X-ray imaging – we expect the transmission of X-rays to reveal the structure and arrangement of bones because the method we use to acquire the image is so similar to our everyday experience of light and shadows. At the opposite end of the spectrum of expectation we find methods such as functional MRI in which we infer altered neural activity by measuring the differential relaxation of an induced nuclear magnetic resonance signal. For fMRI the steps between the investigated biological phenomenon and the observed signal are numerous: increased neural activity – local depletion of oxygen tension – increased blood

R. Bourne, *Fundamentals of Digital Imaging in Medicine*,
DOI 10.1007/978-1-84882-087-6_3, © Springer-Verlag London Limited 2010

flow – increased local oxyhaemoglobin concentration – diamagnetic alteration of spin relaxation rate; and all of this changing on a time scale of seconds. The demands on image processing and the potential for artifacts are considerable. A user of the technology who is unaware of it's assumptions and limitations is likely to misinterpret the images produced.

## 3.2  The Energetics of Imaging

Imaging is the creation of a 2D or 3D *representation* of a physical object by measurement of energy emitted or reflected from the object. For the image to contain information about the object it is necessary for the amount of emitted energy to differ according to the local properties of the object. This change of measured energy provides the image contrast. It is also necessary for the *spatial origin* of measured energy to be known in order for the image to have spatial resolution.

An imaged object can be thought of as a collection of point sources of energy. The energy emitted from a point *always* suffers a degree of spatial dispersion as it travels away from the point. This dispersion leads to an uncertainty in the spatial origin of a measured signal. If the spatial origin of a recorded signal is unknown, due to scatter of emitted energy say, or part of the measured signal does not originate from the imaged object, then spatial and contrast information about the object is partially obscured or corrupted. This spurious signal that contains no information about the imaged object is 'noise'. Unfortunately, there is generally no way to tell which part of a measured signal is information and which part is noise. We may be able to estimate the statistical features of the noise but this merely permits estimation of the *uncertainty* of the measurement, not removal of the noise. The method of signal acquisition must be optimized to reduce the contribution of noise.

Nearly all medical imaging devices measure electromagnetic (EM) radiation – the single obvious exception being ultrasound imaging, which measures high frequency pressure waves. It is perhaps remarkable that outside the range of wavelengths detectable by human vision the human body is relatively transparent, or at least translucent, to electromagnetic radiation. You might conclude that human vision has evolved for the specific purpose of examining *surfaces*, and why not. Surfaces provide a nice summary of objects, they suggest size and shape and, thanks to varying amounts of reflection of different wavelengths, surfaces may appear to have a color that further defines an object. If we *could* see through most surfaces then we would probably live very short lives bumping into things and falling down holes. But there is another fundamental reason that we *do* only see surfaces – visible light is absorbed, *it must be absorbed*, by a biological detector – the rod and cone cells of the retina. We have only one nearly transparent tissue in our bodies – the lens and cornea of the eye. Since it is physically inevitable that visible light will be absorbed or bounce off the surfaces of our body it makes sense that we have evolved a method of measuring some of that light in order to better survive in our environment. We have even evolved a mechanism to deal with the limited range of wavelength sensitivity of our eyes. Infrared electromagnetic radiation can be detected by cells in our skin, and this extra sensor helps protect our tissues from high temperature hazards that our eyes cannot detect.

## 3.2.1 Radio Frequencies

The lowest EM radiation energy levels (radio frequency, or RF) are used for MRI (Fig. 3.1). In MRI pulses of RF energy are transmitted into a body placed in a strong magnetic field. The RF energy disturbs the equilibrium magnetization of hydrogen nuclei, mostly those in the body water, and the reemitted photons are measured as the system relaxes back to equilibrium. Contrast in MRI results from the many different ways the reemitted energy can be affected by the local environment. The RF photons used in MRI barely interact with the electrons that bond atoms and molecules, so MRI has no significant effect on the physiology or structure of the body. The absorbed RF energy that is not reemitted is converted to heat. MRI scanners estimate the rate of heating in the patient and limit the rate of RF power transmission accordingly.

### 3.2.1.1 Microwaves

A little higher in energy than the radio frequency photons used for MRI are microwaves. Microwave EM radiation can be used to probe the magnetic resonance

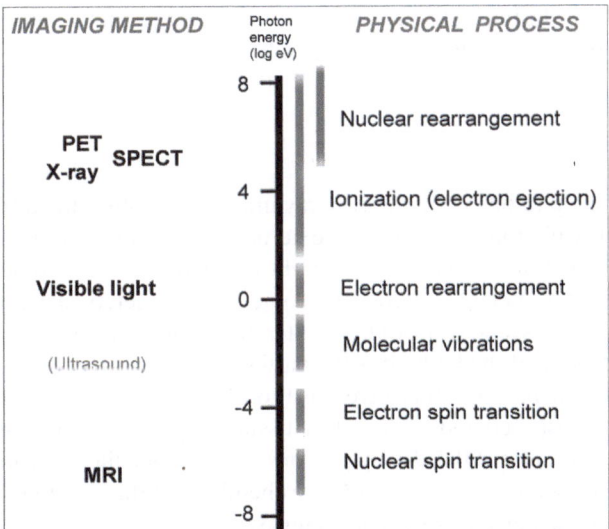

**Fig. 3.1** A logarithmic scale plot of the photon energies used in medical imaging, and the corresponding physical phenomena (Ultrasound imaging utilizes pressure waves, *not* photons, and ultrasound energy is transmitted by *inter*molecular vibrations. All the other imaging methods affect *intra*molecular processes.)

properties of electrons. The phenomenon is called ESR (electron spin resonance) or EPR (electron paramagnetic resonance). Strictly speaking ESR imaging is another kind of MRI. Experimental and analytical biological applications include measurement of free radicals and oxygen tension, however microwaves are not as easy to control or measure as radiowaves and tend to get absorbed by the body without causing the desired electron spin transitions. There are currently no clinical imaging applications of ESR.

### 3.2.1.2   Infrared

The infrared range of photon energies, between microwaves and visible light, can be used to produce temperature images of the human body surface. These *thermograms* have gained limited clinical acceptance for detection of breast cancer.

### 3.2.1.3   Visible Light

Going up the energy scale, at the next medically important EM radiation energy level, the surface of tissue is irradiated with visible light and we measure the photons reflected – this is just photography optimized for medicine. It is called dermoscopy when examining the skin, and endoscopy when looking inside the body. Visible light photons, like RF photons, have negligible effect on the physiology or structure of the body, but they are strongly absorbed and converted into the thermal energy of molecular vibrations. They heat the irradiated surface, but do not penetrate more than a few millimetres below the tissue surface. Visible light is also used routinely in histopathology. In this case very thin sections of fixed and stained tissue are examined microscopically with transmitted light.

### 3.2.1.4   X-Rays

At a much higher photon energy level X-ray imaging measures the differential transmission of X-ray photons by tissue. An external X-ray source irradiates a section of the body with photons – some pass straight through the body, some are scattered and emerge at unpredictable angles, and some are completely absorbed. The unabsorbed X-ray photons, and some of the scattered photons, are measured by a detector on the opposite side of the body. It is the partial transparency of the body to X-rays, or *differential transmission*, that gives rise to a projection image. Although X-ray photons can be absorbed and scattered by tissue, they are not absorbed and reemitted as RF photons are in MRI. X-ray photons cause molecular ionization when they are absorbed and scattered in tissue and the chemistry of the resultant ions can cause permanent tissue damage and genetic changes.

### 3.2.1.5   Gamma Rays

The highest photon energy levels are used in gamma ray imaging. Conventional terminology describes photons emitted by a nuclear decay process as gamma rays, and those emitted by high energy electron collisions as X-rays. There is a small amount of overlap between the highest energy X-ray photons used for medical imaging and the lowest energy gamma photons. The most important distinction between X-ray and gamma ray imaging is that the energy source is *inside* the body in gamma imaging – a radioactive pharmaceutical is introduced into the body and is distributed and localized according to the current physiology and pathology of the body. The two distinct types of radioactive isotope used in medical imaging emit either a gamma photon which can be detected directly, or a positron. A positron travels only about a millimetre through tissue before combining with an electron. The resultant annihilation emits two gamma photons which leave the annihilation site in opposite directions. Whatever the source, emitted gamma photons are detected outside the body by a gamma camera. A plain emission image can be produced or, with a gamma camera that rotates around the body, a tomographic image. Gamma photons, like X-rays, cause ionization in tissue with the result that safety concerns limit the amount of radionuclide that can be administered to a patient. The risk due to the radiopharmaceutical has to be balanced against the need for a signal that will produce an image of adequate clinical quality.

The tissue destructive effects of gamma photons are employed in radiation therapy. A high intensity gamma photon beam produced by a linear accelerator is projected at a target region in the body from multiple directions. The multiple directions serve to maximize the damage to malignant tissue and reduce exposure and damage of surrounding normal structures. The unabsorbed gamma photons are sometimes used to form a low contrast *portal image* of the treatment region. This image is used to confirm the correct targeting of the treatment beam.

### 3.2.1.6   Energy Flow

The medical imaging methods can be crudely categorized according to the flow of energy used for imaging, as illustrated in Fig. 3.2. X-ray and portal imaging use an external energy source and measure differential transmission. Emission imaging introduces an energy source into the imaged object and measures the spatial distribution of emitted energy (differential emission). Thermography measures the spatial distribution of emitted endogenous energy. MR, visible light, and ultrasound all use an external energy source and measure differential 'reflection' of the energy by the body. However, MR is distinct from all of the other imaging methods because it depends on changing the environment (by imposing a strong magnetic field) in order to increase energy absorption to a usable level.

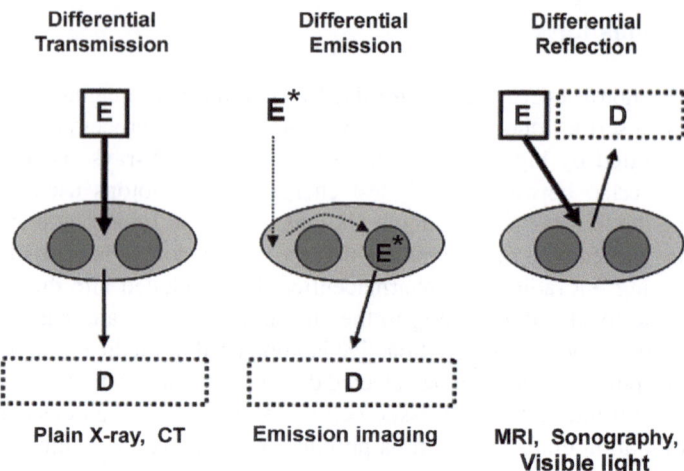

**Fig. 3.2** All imaging techniques measure *energy flow* because energy is essential for information transmission. The common medical imaging methods can be categorized according to the energy flow from source (E) to detector (D). X-ray imaging measures the differential transmission of energy by the imaged object which is positioned between the energy source and the detector. Emission imaging measures the localization of an energy source inside the imaged object – a *latent* photon energy source (E*, a radionuclide) is injected into the body and allowed to distribute according to the specific anatomy and physiology of the subject. The 'reflective' imaging techniques measure differential reemission of energy transmitted from an external source

## 3.3 Spatial and Temporal Resolution of Medical Images

The spatial resolution of a medical image describes the size of the smallest anatomical structures that can be represented independently in the image. The *maximum* spatial resolution of a digital imaging system is limited by the number and spatial distribution of the sensor elements *and* the geometry of the imaging system. However, the *actual* spatial resolution depends on the inherent contrast of an object and the sum of the blurring effects of all the elements in the imaging system including the imaged object itself. These issues are discussed in more detail in Chapter 5.

The idea of 'sensor elements' and system geometry is somewhat different for MRI. In MRI the spatial position of the detected electromagnetic energy does not directly indicate the source or trajectory of the energy as it does in transmission and emission imaging and in sonography. The wavelength of an MRI photon is, in fact, about as long as a patient. The reason we can make a high resolution image with such a poorly localized signal is that we manipulate the local environment with well-defined magnetic field gradients so that the source of the detected energy is encoded in its frequency and phase. The spatial resolution of MRI depends on the strength of the encoding field gradients and the rate of sampling (temporal frequency resolution) of the measured signal – this is, effectively, the *spatial frequency* equivalent of the number and spatial distribution of sensor elements mentioned above. We will discuss the concept of spatial frequency in detail in Chapter 4.

The temporal resolution describes the time required for a single image measurement and the time interval between successive image measurements. For any single image it is usually desirable to acquire the necessary raw data in as short a time as possible. This will reduce the uncertainty in spatial information due to changes in the object that occur during the measurement of the image data. The most obvious example is movement of a patient during imaging, but also important is involuntary movement of internal organs. In functional and dynamic imaging studies multiple images are used to show the time course of spatial changes in the subject – the beating of a heart, for example, or the vascular flow and tissue perfusion of a contrast agent. The actual temporal resolution required depends on the dynamics of the body system studied.

The interrelation between spatial resolution, contrast and noise is illustrated in Fig. 3.3. It is helpful to first discuss the signal and the noise separately.

For any small part of the imaged volume (which may include the object of interest and some air) we can imagine that a specific intensity of 'signal' energy can be detected by the imaging sensor. The amount of signal from each small part will vary from zero to some maximum depending on the way the object interacts with the source of imaging energy. Now consider what happens when we image an 'object' comprised of the two square blocks '$x$' and '$y$' in Fig. 3.3. For each discrete recording (pixel) the imaging system measures the *average* signal intensity in a local region. In each local region the various signals from the parts of the object, and from any background present all add up. There are no negative signals. Where there

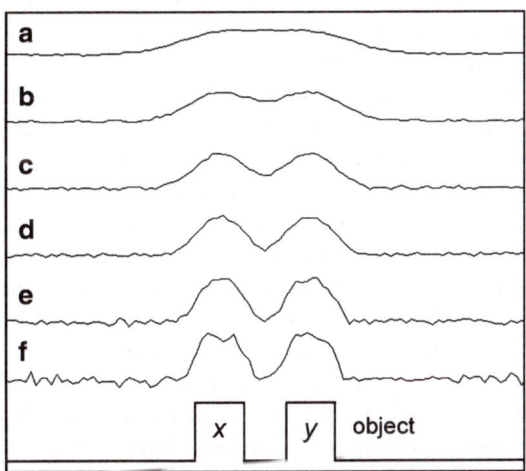

**Fig. 3.3** A 1D representation of the relationship between spatial resolution, noise, and contrast. Profiles **a–f** represent 1D images of a simple object that emits energy from two spatially distinct regions $x$ and $y$. The spatial resolution increases progressively from **a** to **f** but the total signal, the area under the signal profile, remains constant. Contrast, the difference in measured signal intensity from regions $x$ and $y$ relative to the background, increases with spatial resolution but so too does the noise level

is no object present, as in the gap between the blocks '$x$' and '$y$', or outside the blocks, there is no object signal. All the signals are characteristic of the object and the space around it and, so long as the object does not change or move, the signals do not change.

If the spatial resolution of the system is low then the measured local regions are large. If the spatial resolution of the system is high then the local regions are small. The important differences here are: (1) When the measured local regions are small the total measured signal for each region is small, and (2) When the measured local regions are small the variations in measured signal *between* local regions more closely reflect the actual spatial distribution of energy. In other words, higher spatial resolution means *less* signal in each measurement region (pixel).

Now let's ignore the signal from the object and consider the noise that is inevitably part of the measurement. We can imagine, somewhat simplistically, that *the same* amount of random noise is present in *all* regions of the image field. If we just measured the noise for a *long* period of time then the measurements in each region (pixel) would be the same and they would be due solely to noise. In reality, however, we don't (or can't) make measurements for a very long time. Because the noise is random the *actual* contribution of noise to the measurement is different in each pixel. The relative differences get bigger as the measurement time gets shorter. Also, the relative differences get bigger as the measurement volume gets smaller, because there is less spatial averaging of the random noise. Higher spatial resolution means *greater* differences between the noise contributions to each measurement region (pixel).

Now we can put our considerations of signal and noise together. At low spatial resolution (profile a in Fig. 3.3) we have a relatively strong signal for each pixel and all pixels have a relatively constant noise contribution. However, this was achieved by spatial averaging and leads to a relatively imprecise (blurred) representation of the spatial variations of object contrast. At high spatial resolution (profile f in Fig. 3.3) we have a relatively weak object signal for each pixel and a relatively large difference between the noise contributions to each measurement. However, there is less spatial averaging so the signal part of the measurement more closely represents the spatial variations of object contrast. Unfortunately, this improvement in spatial resolution of contrast detail is partially offset by the larger random variations in measurements due to noise. Overall we can say that the *Signal-to-Noise Ratio* (SNR) *decreases* with increasing spatial resolution.

It is generally the case that a longer measurement time will improve the information content (contrast) of an image relative to the amount of noise – in other words, the SNR will be higher. If a longer measurement time is undesirable, as in the case of actual or potential subject movement or a need for high temporal resolution, then measurement with a more intense energy source will also increase the SNR. In the case of X-rays and gamma rays this means a higher risk to the patient.

The choice of imaging method for a particular investigation depends on the size of relevant anatomical features (spatial resolution required), the specificity of information required (the type of contrast) and, in functional investigations, the time scale of relevant changes (temporal resolution). A comparison of the approximate

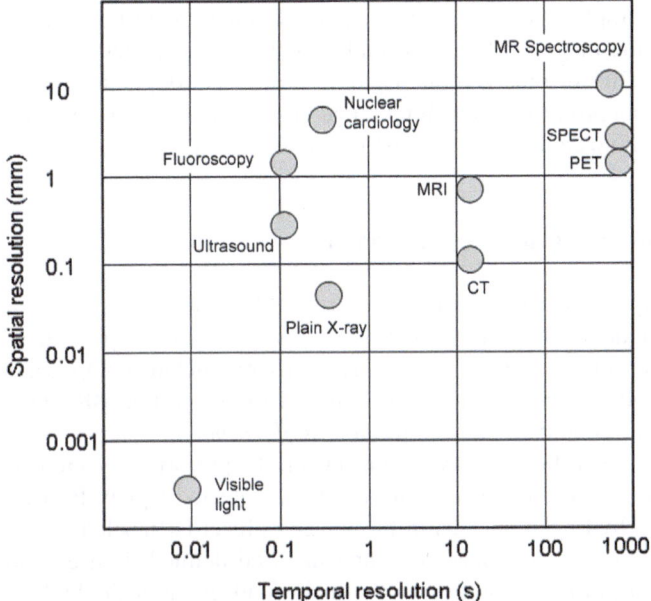

**Fig. 3.4** Typical spatial and temporal resolution of common medical imaging techniques. Spatial resolution is limited by the dispersion of the signal during and prior to detection. Temporal resolution is limited by the strength of the signal. The application of the method, whether structural or functional, depends on both spatial and temporal resolution, and the type of contrast detected

spatial and temporal resolution performance of the major medical imaging methods is illustrated in Fig. 3.4. Notice that this plot does not describe the type of contrast available from each method.

## 3.4 Medical Imaging Methods

We now look more specifically at the way images are produced by each of the common medical imaging modalities. Once again we will start with the least energetic photons and work our way up the electromagnetic energy scale before jumping to kinetic energy – ultrasound.

### 3.4.1 Magnetic Resonance

The phenomenon of nuclear magnetic resonance was for a long time exploited almost exclusively by organic chemists for the elucidation of molecular structures. The methods depended on the ability to create a highly homogeneous magnetic field in the sample, but it was eventually realized that by introducing a well defined gradient in the magnetic field it would be possible to spatially map the nuclei in the sample. This idea marked the beginnings of MRI. Magnetic resonance imaging techniques can probe an astonishing range of structural and functional properties of the body with no significant effect on physiology.

### 3.4.1.1 The $^1$H Nucleus: Water and Fat

The majority of MRI techniques measure the tiny signal arising from transitions between magnetic energy levels (associated with different 'spin states') of the hydrogen nuclei in water. Because the $^1$H hydrogen nucleus comprises just a single proton $^1$H MR techniques are often referred to as 'proton MR' even though all atomic nuclei contain protons. Proton MR *does not* measure a signal from any protons other than those in hydrogen nuclei. To perform any kind of imaging it is necessary to know the spatial origin of the detected signal. In MRI the spatial position of a hydrogen nucleus in the imaged object is 'labeled' with a particular resonance frequency and phase by imposing well-defined short duration magnetic field gradient pulses in addition to the main field. Even at the highest achievable spatial resolution the measured MR signal for each voxel is the average signal from a collection of billions of nuclei.

Two factors make the MRI signal very weak. Firstly, the energy involved in a transition between nuclear $^1$H magnetic energy ('spin') states is ten orders of magnitude less than that of an X-ray photon (Fig. 3.1). Secondly, detectable transitions arise only due to the *very slight* excess of nuclei in the lower energy state. This excess depends on both the temperature and the magnetic field strength, and is only $\approx 0.0005\%$ at body temperature in a 1.5 Tesla MRI magnet. Fortunately there are plenty of hydrogen nuclei available in biological tissue: $\approx 5 \times 10^{22}$ in a cubic centimeter. The energy difference between spin states is proportional to the magnetic field strength. These two factors, population and energy difference, are the reason for the gradual replacement of 1.5 T MRI scanners by 3 T and higher field systems – they produce more signal.

In the absence of the MRI magnet, only the Earth's magnetic field ($\approx 0.00005$ T) would be present. Both the population and energy differences between the spin states would be very much smaller and no clinically useful signal would be available. Nevertheless, using a small 'prepolarizing' electromagnet to boost the population difference, it is possible to perform low resolution MRI in the Earth's field (Fig. 3.5).

In biological tissue the next most common hydrogen nuclei after water are those present in fat or lipid, especially the $-CH_2-$ groups that form the backbone of fatty acids. In general, the fat signal does not contain any diagnostically useful information, and because the hydrogen nuclei in lipid resonate at a slightly different frequency from those in water, the part of an MR image arising from lipid may be displaced relative to the adjacent tissue water image. This is a *chemical shift artifact*. It can be avoided by suppression of the lipid signal during the MRI measurement.

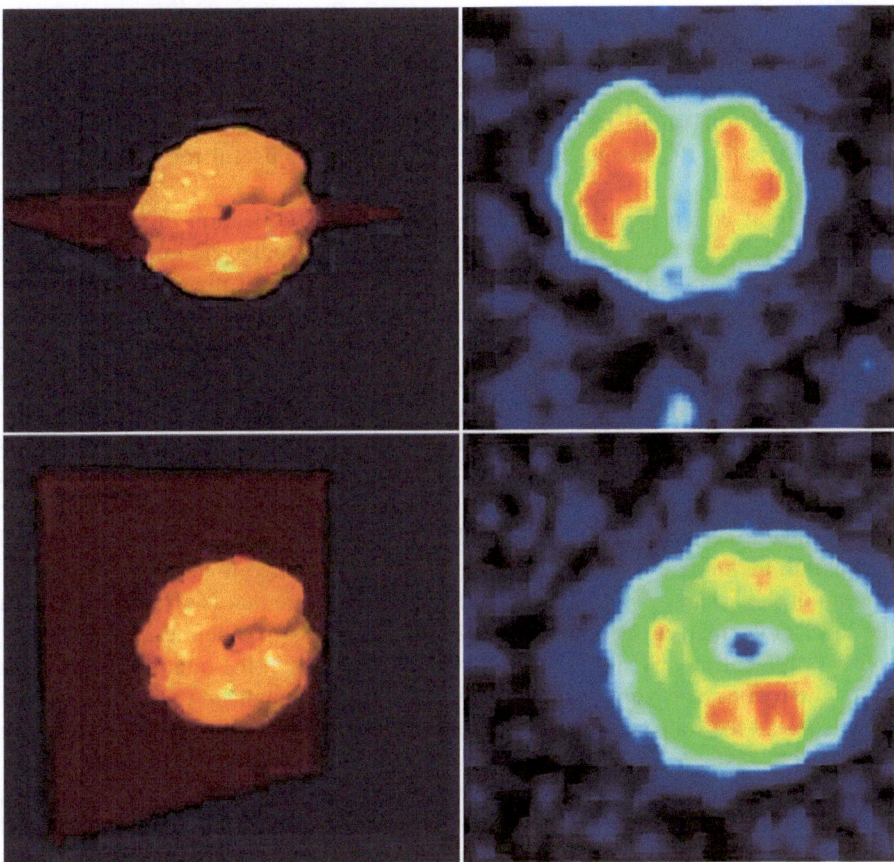

**Fig. 3.5** Earth's field MRI of a mandarin. Medical MRI depends on measurement in a strong magnetic field in order to acquire adequate signal strength. However, low resolution (and low cost!) MRI is possible even in the very weak ($\approx 0.00005$ T) magnetic field of the earth (image courtesy of Dr. Andrew Coy, Magritek Limited, Wellington, New Zealand. www.magritek.com)

There are many different ways of generating contrast in MRI, each depending on a particular physical property, or mixture of physical properties, of tissue. All MRI techniques depend on input of RF energy, manipulation and evolution of the stimulated system, and measurement of the RF energy emitted as the system relaxes back to thermal equilibrium. In MRI the term 'thermal equilibrium' refers to the populations of nuclear spin states that exist at a particular temperature and magnetic field strength. Absorption of RF energy disturbs this equilibrium. Emission of RF and other energy transfer processes restore the equilibrium. Disturbance and restoration of thermal equilibrium, at least in MR measurements, does not mean the temperature of the sample is going up and down significantly. However, the small proportion of absorbed RF energy that is converted to heat *can* cause significant tissue heating in RF-intensive MRI pulse sequences.

The simplest form of MRI contrast is dependent on the amount of water present in a particular tissue or, strictly speaking, the water hydrogen nucleus density. These are called *proton density* (PD) images. Since the range of proton densities in tissue varies only in the range 70–100% compared with pure water not much contrast can be generated. PD images are, however, particularly useful for examination of cartilage.

The rate at which the *measurable* signal from this collection of nuclei decays depends on a large number of factors in the local molecular environment of each nucleus. If the 'system' (the collection of $^1H$ nuclei) is not manipulated in some way, by application of extra magnetic fields or extra RF energy, then the measurable RF signal will generally decay to zero within one second of input of the first pulse of RF energy. Differences in this rate of signal decay ($T_2$ *relaxation*) provide one method of MR contrast generation ($T_2$ contrast).

The lack of a measurable RF signal does not mean that a stimulated system has relaxed back to thermal equilibrium. It can simply mean that the emitted signals are incoherent – they cancel each other out. To a limited extent the system can be manipulated to recover some coherence but ultimately no RF signal can be detected because all the input energy has been reemitted or converted to heat. Thermal equilibrium of the system is achieved ($T_1$ *relaxation*) at a rate that is also dependent on the local molecular environment, though due to factors not identical to those that cause $T_2$ relaxation and signal incoherence. Differences in the rate of equilibration of local environments can also be used as a contrast mechanism ($T_1$ contrast). The time required for water protons in biological material to reach thermal equilibrium is typically 5–10 s.

Because the position of nuclei in MRI is labeled with their frequency and phase, nuclei that change position in the imaged object between the time of labeling and the time of measurement may give a 'spurious' or 'incorrect' signal. Such molecular movements might be due to fluid flow, simple diffusion, or patient movement. In the case of patient movement the result is often a distinctive *movement artifact* evident in the image. Flow and diffusion, on the other hand, can cause a signal increase or decrease depending on whether signal coherence is lost or gained. Diffusion-weighted and perfusion-weighted imaging methods develop contrast according to the physical mobility of water in tissue and body fluids.

The structural components of cells and tissues affect the microscopic and macroscopic molecular environment and thus the MR signal from water indirectly reflects both body structure and function. If the tissues of interest in a particular examination do not have good inherent contrast then the local molecular environment can sometimes be modified by administration of MR contrast agents. These are mostly based on paramagnetic metals that affect the $T_1$ and $T_2$ relaxation rates.

### 3.4.1.2 Metabolites: MR Spectroscopy

All the mobile hydrogen nuclei in tissue, not just those in water, give rise to proton MR signals. The signal from protons in metabolites present at millimolar concentrations is roughly 10,000 times weaker than the signals from the ≈50 molar water. With efficient suppression of the water signal it is possible to measure the signal from a few metabolites and thus create functional metabolic images. For a metabolite to be *MR visible* it must be small, freely mobile (not bound tightly to a large molecule such as a protein), and present at millimolar concentration. Because most metabolites contain multiple non-identical hydrogen environments each compound has multiple resonances and these may not be distinct from those arising from other metabolites. In order for *any* of these signals to be distinguished from each other, and thus for any specific metabolite to be identified, we need much greater magnetic field homogeneity than is required for water imaging. This is a major problem for in-vivo spectroscopy – especially outside the brain, and in brain regions close to the sinuses or bone. Under ideal conditions it is possible to spatially map, in two or three dimensions, the relative intensities of some metabolite resonances. The highest useful spatial resolution achievable in a human is about 0.5 cc. These *Chemical Shift Images* (CSI), or Spectroscopic Images *roughly* represent the spatial distribution of the metabolites (Fig. 8.10).

### 3.4.1.3 Other Nuclei

The success of MRI is largely due to the high concentration of water in soft tissue, and subtle but detectable differences in relaxation properties of the water proton depending on tissue type. The MR signal of other biologically important nuclei, including phosphorus, nitrogen, sodium, and carbon, can also be measured, but their signals are even weaker than those of hydrogen nuclei. Because each nucleus resonates at a specific frequency in a given magnetic field, dedicated and expensive hardware is required for each nucleus. The nuclei other than hydrogen are mainly used for studies of physiology rather than tissue structure and, so far, mainly in research. A few examples include helium and xenon gas for imaging lung function, phosphorus for bioenergetic studies, and carbon and nitrogen for metabolic studies.

## 3.4.2 Visible Light Imaging

Visible light interacts strongly with tissue and thus does not penetrate more than a few millimetres into the body, the single exception being penetration into the eye. Visible light imaging is therefore used for characterization of the body surface, the internal surfaces of the eye, and the body cavities accessible with an endoscopic camera. In the latter case, because light does not penetrate deep into the body, we have to introduce a light source together with the camera. (To be precise, it is actually a flexible *light guide* that is introduced into the body. The camera and light source remain outside the body.)

At first glance we might think of dermoscopy, skin surface imaging, as an obvious exception to the generalization that medical images represent physical properties of

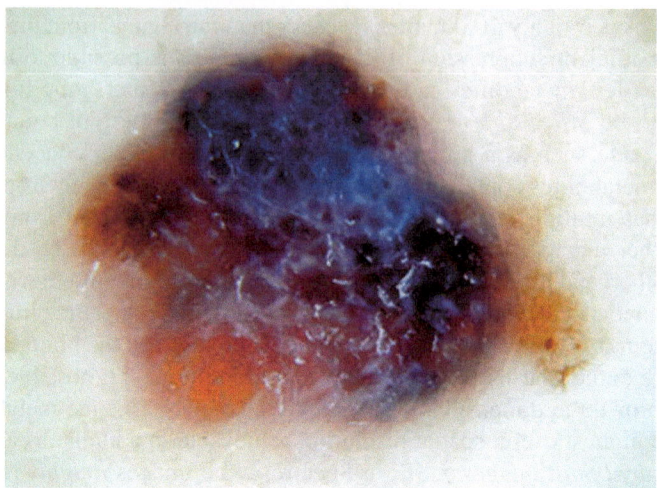

**Fig. 3.6** Dermoscopy uses visible light to image the body surface – in this example a skin lesion. Cross-polarization limits reflection of visible light from the stratum corneum in order to make the epidermis transparent and thus to reveal morphological features of the subdermal layers that are not visible to the naked eye (image courtesy of 3Gen, LLC, San Juan Capistrano, CA, USA)

tissue that are invisible to the human eye. In fact dermoscopy techniques specifically aim to limit reflection of visible light from the stratum corneum in order to make the epidermis transparent and thus to reveal morphological features of the subdermal layers that are *not* visible to the naked eye. This can be achieved by a layer of oil or other liquid between the tissue and the camera lens, or by illumination of the skin with polarized light (Fig. 3.6).

### 3.4.3 X-Ray Imaging

X-ray imaging is used for both structural and functional studies. *Differential transmission* of X-ray photons by body tissue is fundamental to image formation – it is the mechanism of contrast generation. There are two predominant mechanisms by which X-ray photons interact with body water and tissue: Compton scattering, and the photoelectric effect. In the absence of Compton scattering we would get very nice X-ray images because all photons would either be absorbed by the tissue according to its atomic composition, density and thickness, or they would pass straight through the tissue in a straight line from source to detector. Unfortunately most X-ray photons bounce or glance off the atoms in tissue and change direction without absorption – the Compton effect. Many of these photons reach the detector and contribute to the measurement but because their trajectory is unknown (and unknowable) they contribute no useful spatial information about the subject, only noise. Other Compton scattered photons shoot randomly out of the patient and create a radiation hazard.

### 3.4.3.1 Plain X-Ray

The most common form of X-ray imaging is plain projection. A conical beam of X-rays is directed at the body and constrained (collimated) to the region of interest. The X-ray photons are differentially absorbed and scattered by the body according to the types and amounts of tissue they encounter. The photons that pass through the body unabsorbed and unscattered form the diagnostic image on a planar image receptor behind the body. Scattered photons that reach the image receptor are of indeterminate spatial origin and thus only add noise and diminish the contrast of the image. The image formed is a *projection image* (Fig. 3.7).

The image receptor, or detector, was traditionally a silver halide film sensitized with a fluorescent screen that converts single X-ray photons to very large numbers of visible light photons. More recently, photochemical screen/film systems have been partially replaced with 'computed radiography' (CR) systems comprised of a photostimulable phosphor plate that can be 'read' with a laser and converted to a digital image. CR systems obviate the need for wet chemistry film processing and the laser reading system produces a digital image output – with its associated advantages. More recently still, flat panel semiconductors that directly detect X-rays have been developed. The image data is read directly off the sensor without need for laser scanning or phosphor regeneration as is required for CR detectors.

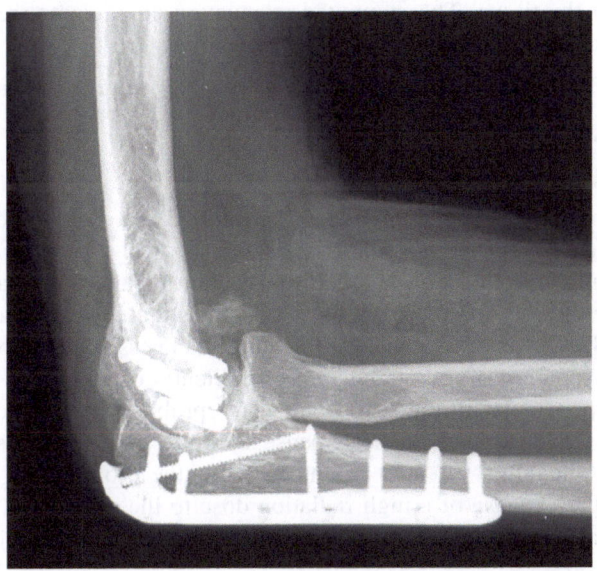

**Fig. 3.7** Plain projection X-ray of a surgically reconstructed elbow. The signal intensity at any point in the image represents the sum of all the attenuation coefficients of the tissue, metal, and any other matter that lie on the line between X-ray source and detector, *plus* any scattered energy. Because the attenuation coefficient of steel is much higher than that of any human tissue, no bone or soft tissue detail is visible where surgical steel is present

In a plain projection image here is no resolution of separate structures that lie in the same line between the X-ray source and the image receptor. All these struc- tures contribute, according to their individual X-ray attenuation coefficients, to the contrast developed at a single point in the image. The projection image can thus be thought of as a 2D summary or average of the 3D body of tissue (and any other matter) that lies between the X-ray source and the image receptor. For many pur- poses this lack of depth resolution is not a major problem. It is partly obviated by routinely acquiring further projection images of the same anatomical region with a different, roughly perpendicular, projection axis. In this way structures not resolved due to overlap in one projection may be seen separately in another view.

In most cases we can say that X-ray imaging is *contrast limited* and this is due to the lack of difference in X-ray attenuation of different tissue types. Only bone and teeth are distinctly different from other tissues. The visibility of some structures can be enhanced by introduction of a specific contrast agent into the body.

### 3.4.3.2   Computed Tomography

Tomography is the creation of sectional images of an object. X-ray computed to- mography (X-ray CT, or just CT) uses multiple projection images to construct a cross sectional image of the body that represents a 2D map of the X-ray attenuation coefficients of the tissue. The computed attenuation coefficients are stored as *CT numbers* which represent the attenuation relative to water:

$$CT = \frac{\mu - \mu_{water}}{\mu_{water}} K \qquad (3.1)$$

where $\mu$ is the linear attenuation coefficient and $K$ is a scaling constant. When $K = 1{,}000$ the CT numbers are *Hounsfield Units*.

This 2D map is distinct from the 2D projection of a plain X-ray. In a tomo- graphic image there is no superposition of structures that are coplanar in the image slice (Fig. 3.8). However, the slice has a finite thickness so there is effective su- perposition (averaging of the attenuation coefficients) of structures that lie in the same line perpendicular to the slice plane. This problem is reduced by minimiz- ing the slice thickness, but at the expense of SNR. Exactly the same problem applies to MR imaging. CT gets around many of the problems of projection ra- diography at the expense of a high radiation dose to the patient. In the developed world half an average human's lifetime ionizing radiation dose now comes from CT imaging.

When there is a need for plain projection images these can be synthesized ('re- constructed') from 2D or 3D CT data. This is a *Digitally Reconstructed Radiograph* (DRR). With a 3D CT data set a DRR can be created for any arbitrary projection.

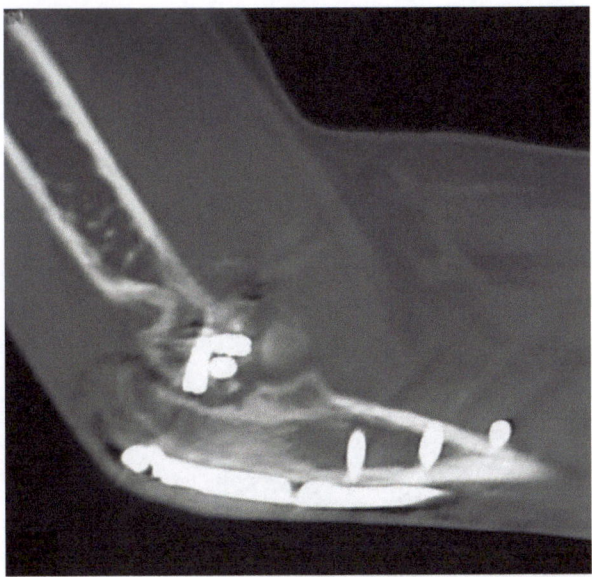

**Fig. 3.8** A CT X-ray image of the same elbow shown in Fig. 3.7. The image represents a *thin section* through the joint. Notice that only parts of the surgical metal are visible – the parts that lie in the plane of the section. There is poor soft tissue contrast in this image because a bone window has been applied

This is particularly valuable for radiation therapy planning where it is desirable to orient the treatment beam along a series of lines that will obtain maximum treatment effect while at the same time minimizing the dose to normal tissue (Fig. 3.13).

### 3.4.3.3 Fluoroscopy

Fluoroscopy is a general term used to describe X-ray imaging techniques that utilize an *image intensifier* to produce images from low intensity X-ray beams. The image intensifier converts a small number of X-ray photons to a very large number of visible light photons – far more than are generated in a conventional fluorescent screen – and these are detected by a video or digital camera. The purpose of fluoroscopy is to display images semi-continuously in real time in order to perform functional studies, and to guide surgery and interventional procedures (Fig. 3.9). The main requirement is high temporal resolution and merely *adequate* spatial resolution and contrast. Modern digital fluoroscopy does not use a continuous X-ray beam. Instead a series of short exposures are made at a time interval, the *frame rate*, that is adequate to represent significant features of the procedure. The quality of the images depends on the intensity of the individual exposures.

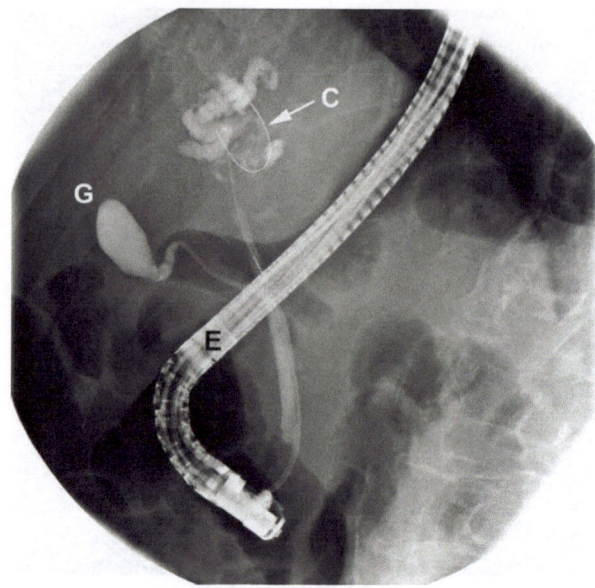

**Fig. 3.9** Fluoroscopic image acquired during endoscopic retrograde cholangiopancreatography (ERCP). A catheter (C) is introduced into the bile duct from the duodenum via optical guidance from the endoscope (E). Fluoroscopy provides multiple images for guidance of the catheter up the bile duct and into the liver. Contrast agent has been injected, via the catheter, into the bile duct, filling the gall bladder (G) and ducts in the liver

### 3.4.4 Emission Imaging

Emission imaging differs fundamentally from X-ray imaging in the source of the measured radiation. In emission imaging a radiation source is injected into the body and an external gamma ray detector (gamma camera) is used to spatially map radiation exiting the body and thus deduce the internal distribution of the radiation source. The emitted radiation has an energy specific to the particular nuclide (radioisotope) used. As for X-ray imaging, scattering of radiation blurs the acquired image because the source of detected scatter photons is unknown. However, in emission imaging, absorption of radiation by the body is a nuisance as it reduces the useful detected signal.

Contrast in emission imaging does not result from differential absorption of the imaging energy by body tissue. Instead contrast depends on differential or specific localization of the energy source within the body. This specificity can be matched to particular functions (e.g. heart muscle perfusion) by attachment of the nuclide to a specific pharmaceutical which is thus, effectively, a biochemical contrast agent.

Because radiation absorption and scattering are always present, image reconstruction methods often attempt to adjust the image according to expected absorption patterns. This *attenuation correction* depends on calculation of the *expected* attenuation from an anatomical image acquired by another imaging method such as

CT or MRI. Alternatively, attenuation can be estimated by scanning the body with a *line source* and obtaining a transmission image.

Emission images produced without attenuation correction generally have very poor spatial resolution. The need for attenuation correction and correlation of emission image data with higher resolution anatomical images has led to the development of combined scanners, both SPECT-CT and PET-CT. The ability to acquire both image types without changing the patient position and without a significant time interval provides a major improvement in quality of attenuation correction and correlation of clinical information.

One method of minimizing the effect of scatter radiation is to measure, not just the spatial origin of detected photons, but also their energy. Scattered photons have lost some energy in collisions so, provided that the detector can measure photon energy with adequate precision, those photons that have less energy than the characteristic energy of the nuclide can be ignored.

Gamma ray photons in medical imaging are usually detected with an *Anger Camera* (a modified version in the PET detector) by a method similar *in principle* to the phosphor-mediated detection of X-rays. Single high energy photons are first converted to multiple visible light photons which are more easily and more reliably measured with conventional detectors. However, high energy gamma ray photons are less likely to interact with the phosphor screens used for X-rays (and extremely unlikely to interact with a direct semiconductor detector). Instead a thick (several centimeters) fluorescent crystal (scintillant) is used to convert the gamma ray photons to visible light. The thickness and density of the crystal ensures there is a high probability of photon conversion, however, it also leads to considerable lateral dispersion of the emitted visible light. The Anger camera corrects this problem by detecting the emitted light with a close-packed array of photomultiplier (PM) tubes. Each converted gamma photon produces visible light photons that disperse laterally in the scintillant crystal and produce an output in a group of adjacent PM tubes. The location of the original photon conversion can be determined by comparison of the relative outputs of the PM tubes. Summation of the outputs of all the PM tubes gives an indication of the gamma photon energy. This *energy discrimination* permits rejection of counts due to reduced energy scatter photons.

Single photon emission computed tomography (SPECT) uses one or several moving gamma cameras to obtain a series of plain emission images from which a tomographic image can be constructed. Figure 3.10 shows images from a $^{67}$Ga-citrate SPECT study. This radiopharmaceutical is commonly used for tumor and inflammation detection. In this example there is considerable non-specific tracer uptake in the skeleton but the major activity demonstrates inflammatory tissue in and around the right kidney.

Positron emission tomography (PET) imaging is based on the physiological localization of introduced positron-emitting radionuclides. Because PET detects pairs of photons known to have essentially colinear trajectories, the spatial localization of the radiation source can be assigned to a *line of response* (LOR) as illustrated in Fig. 3.11. Modern PET detectors can measure both the photon energy and the time difference between detection of photon pairs. Precise measurement of photon energies permits the rejection of signals from the lower energy scattered photons that

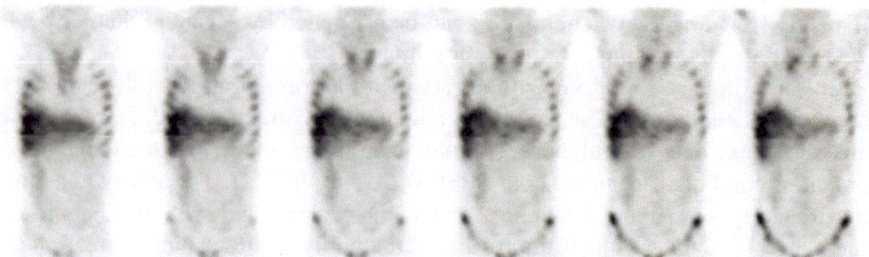

**Fig. 3.10** Typical image data from a SPECT scanner. These coronal $^{67}$Ga-citrate images (slices derived from a 3D data set) demonstrate an infection following renal transplant. Note the strong accumulation of this tracer in bone (image courtesy of Department of Nuclear Medicine, PET and Ultrasound, Westmead Hospital, Sydney)

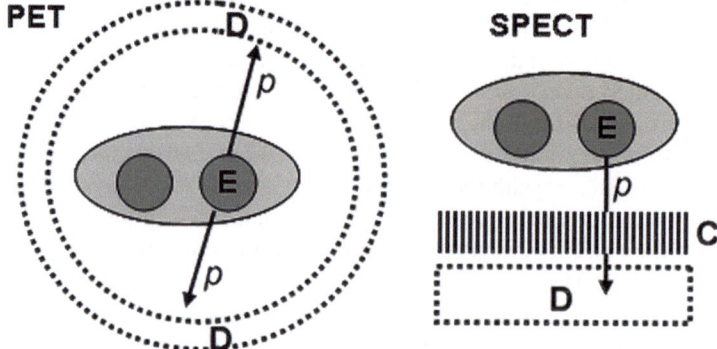

**Fig. 3.11** Comparison of photon detection in PET and SPECT. In PET a pair of photons ($p$) with collinear trajectories is detected and the source of radiation can be localized to a *line of response* (LOR) inside the detector ring. Signal localization and scatter suppression in SPECT and planar emission imaging requires a collimator (C) that inevitably reduces sensitivity due to absorption by the collimator of some unscattered 'signal' photons

have an undefined spatial origin. The time difference between detection of photon pairs permits the localization of the radiation event to a specific region on the LOR. This is a major advantage of PET over SPECT.

One of the limitations of PET is the relatively short half life of positron emitting radionuclides (e.g. 109 minutes for the most common nuclide $^{18}$F) compared with the single photon emitters used in SPECT. Some of the short-lived nuclides must be prepared in a cyclotron located at the same site as the PET scanner.

Unlike the large metal nuclei used in single photon gamma imaging, the common PET nuclides ($^{15}O$, $^{13}N$, and $^{11}C$) are normal constituents of biomolecules. The other common PET nuclide ,$^{18}F$, is biochemically very similar to the hydrogen for which it substitutes. Thus PET is useful for direct studies of metabolism when the nuclide is substituted into a specific biomolucule. Figure 3.12 shows images from a PET-CT investigation of a patient with malignant melanoma. Compared with the

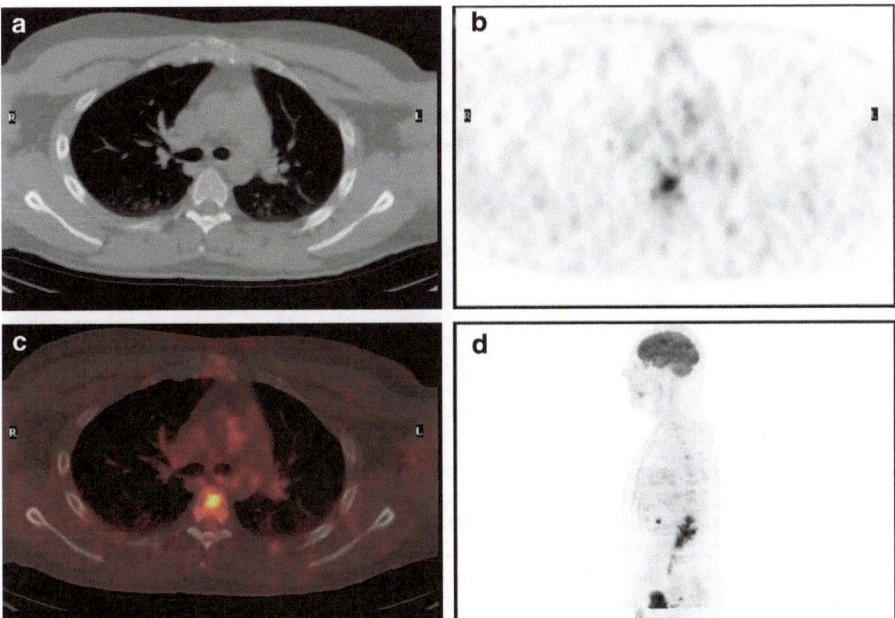

**Fig. 3.12** Typical image data from a PET-CT scanner. This FDG ($^{18}$F-deoxyglucose) scan demonstrates melanoma metastatic to the spine. The transverse CT image (**a**) shows no unusual signs in the vertebra. In contrast, the corresponding transverse PET slice (**b**) shows distinct focal $^{18}$F accumulation, although its anatomical location is poorly defined. Fusion (coregistration) of the PET image with the CT section (**c**) confirms the vertebra as the site of FDG localization. Note the use of a color map to distinguish the PET data from the CT data in the fused image. The saggital MIP (maximum intensity projection) (**d**) is a synthesized projection image, analagous to the CT-derived DRR mentioned above (image courtesy of The Department of Nuclear Medicine, The Blackrock Clinic, Dublin, Ireland)

SPECT images of Fig. 3.10 the emission image has relatively high spatial resolution. However, without the complimentary CT image the precise anatomical location of the detected hot spot (localized volume of high emission activity) would remain uncertain.

Emission tomography images are often reviewed by physicians as movies that are rotating projections of a 3D data set. The moving image often enhances the visibility of hot spots that are difficult to detect in a series of static projection or slice images.

### 3.4.5 Portal Images

Portal images are formed by a gamma ray beam used for radiation therapy. Due to the very small difference in gamma ray attenuation between bone and soft tissue, portal images have very poor contrast. Patient and target movement is to some

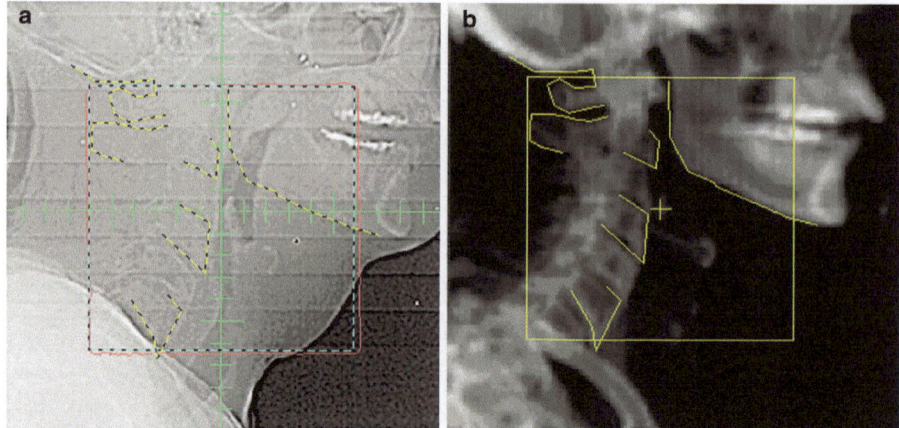

**Fig. 3.13** Different imaging methods provide complimentary information. Here a megavoltage portal image (**a**) produced during radiation therapy of a head and neck malignancy, is compared with a separately acquired X-ray projection (DRR from CT data) of the same region (**b**). Note the relatively poor contrast in the portal image, especially between bone and soft tissue, and high noise level. Correct targeting of the treatment beam can be confirmed by alignment of landmarks (yellow lines) defined in the high contrast X-ray image with the same anatomy in the portal image (from Lee, T, 2008, 'Evaluation of an intensity modulated radiotherapy (IMRT) cast system for head and neck cancers' Masters (Research) thesis, University of Sydney, with permission)

degree inevitable during radiation treatment and, since most treatments require doses spaced some days apart, there may be differences in patient positioning at each session. For quality assurance it is important to assess the degree to which the actual delivered treatment matches the treatment plan. The purpose of the portal image is to acquire, with the patient in the current treatment position, an anatomical image with a short exposure of an uncollimated treatment beam. This image can be compared with the original high contrast planning image (typically a DRR from CT) to assess any errors in treatment beam position, and if necessary adjust subsequent treatment sessions accordingly (Fig. 3.13).

### 3.4.6 Ultrasonography

Sonography is an odd man out in medical imaging as it is one of very few methods that *does not* employ electromagnetic radiation as the imaging energy. Instead sonography uses high frequency pressure waves and thus affects and measures the total kinetic energy of the molecules in tissue rather than the internal energy of their electrons or nuclei. The pressure waves are generated in very short (microsecond) bursts by an external transducer applied directly or indirectly to a tissue surface – most commonly the skin. The majority (99%) of the operational time the transducer is used as a sensor to detect the pressure waves that are reflected by tissue interfaces of dissimilar acoustic impedance.

Sound (pressure) waves are attenuated in tissue in inverse proportion to their wavelength. The highest frequencies (15–20 MHz) are completely attenuated within a few millimeters in soft tissue, while the lowest frequencies (1 MHz) can traverse the torso of a large human or halfway through an elephant.

Since the speed of sound in tissue is roughly constant, the time at which echoes are detected indicates the distance from the transducer to the surface at which the echo was formed. This information, and the ability to direct the transmitted pulses in a specific direction via a multi-element transducer, provides enough information to construct an image from the measured echoes. The image intensity is a display of the intensity of the echoes – the greater the impedance difference the more intense is the reflected pressure wave.

The strong absorption and scattering of sound waves in tissue mean that ultrasound imaging has quite poor SNR in comparison with the other common medical imaging methods (Fig. 3.14). However, ultrasound has many attractive advantages: very high temporal resolution; good soft tissue contrast; minimal biological hazard; relatively low cost; and portability to name a few.

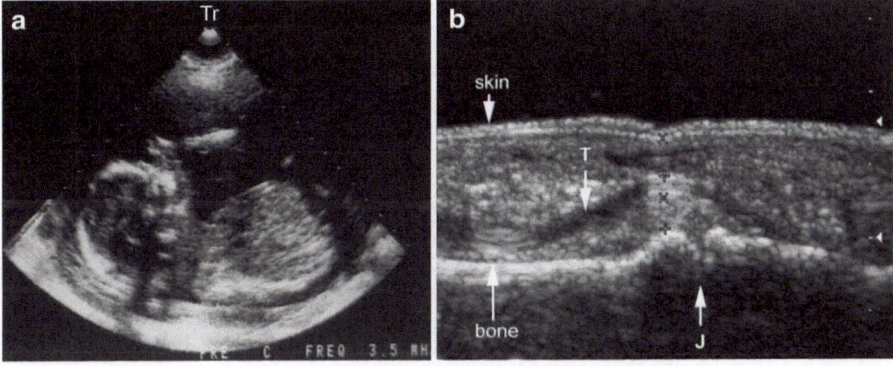

**Fig. 3.14** Ultrasound images of the author's nephew (**a**), and the palmar surface of a finger immersed in water (**b**). The fetal scan (**a**) illustrates the typical sector image produced by a transducer (Tr) that transmits a 'sweeping' beam of pulses. In image **b** the transducer is about 2 cm from the skin surface so the sector pattern is not visible. The signal intensity is highest where the reflected sound wave intensity is highest – at the interface between regions of distinctly different acoustic impedance. In image **b** the signal beyond the bone is weak because most of the sound energy is reflected by the bone, except at the joint space (J). The flexor tendon (T) appears dark (hypoechoic) because of its slightly higher density than surrounding soft tissue. In general the noise level increases with depth due to attenuation (note that the ultrasound image fails to show the prodigious mathematical talent latent in the fetus)

## 3.5 Summary

- Medical imaging produces images of body structure and function by measurement of the differential flow of energy. Most medical imaging depends on the tissue-penetrating properties of energy that cannot be directly detected by the human senses – especially electromagnetic radiation outside the visible light spectrum, and high frequency sound waves.
- Medical imaging with electromagnetic radiation depends on the interaction of photons with body tissue. Photons are absorbed, scattered, and emitted by tissue according to the specific atomic and molecular composition of the tissue at the time of imaging. The nature of the interactions between the imaging photons and the tissue is dependent on the energy of the photons. The *image contrast* thus depends on both the photon energy and the tissue composition.
- Ultrasound imaging depends on the differential transmission and reflection of molecular kinetic energy. The primary mechanism of image contrast formation is the reflection of sound waves at the boundaries between tissues of distinct acoustic impedance.
- Formation of image information depends on an imaging system's ability to precisely and accurately measure both the *intensity* and *spatial localization* of the imaging energy. The choice of imaging modality depends on the type of tissue contrast required (the structures or functions of interest), and on the required spatial and temporal resolution.
- The practically achievable spatial and temporal resolution is often limited by the inherently weak imaging energy flow relative to background noise (as in MRI), the need to limit imaging energy flow due to its biologically adverse effects (as in X-ray and emission imaging), and by scattering of the imaging energy and consequent loss of spatial information (as in X-ray, emission imaging, and ultrasound).

# Chapter 4
# The Spatial and Frequency Domains

## 4.1 Introduction

The interconversion between spatial and frequency domains using Fourier and other transforms is of critical importance in image processing and, for some imaging methods, the construction of images from raw scan data. A significant feature of the transforms is that we can convert back and forth between spatial and frequency domains without loss of information or introduction of noise.

## 4.2 Images in the Spatial and Frequency Domains

### 4.2.1 The Spatial Domain

The concept of the spatial domain requires little introduction. We live in it! Most of the images we are familiar with are spatial domain images — they display a matrix of color or gray scale intensities in a 2D spatial plane. They represent a discrete sampling of the change in intensity of a signal in space and there is a direct correspondence between the coordinates in the image and space in the 'real world'.

We can perform image processing operations directly on these spatial domain images and we often do. Most domestic image processing software, for example Adobe Photoshop, operates exclusively in the spatial domain. However, there are image adjustments that are faster and more precise if we perform them after first transforming the spatial domain image into its frequency domain equivalent. Also there are some image adjustments that can *only* be performed in the frequency domain. Sometimes we acquire raw image data in the spatial frequency domain, most notably in MRI, and it must be converted into the spatial domain in order to create an interpretable anatomical image.

R. Bourne, *Fundamentals of Digital Imaging in Medicine*,
DOI 10.1007/978-1-84882-087-6_4, © Springer-Verlag London Limited 2010

## 4.2.2 Common All-Garden Temporal Frequency

When we encounter the term *frequency* we usually think about regular oscillations. Some examples might include radio waves, audio waves, an ultrasound signal, or perhaps the waves at the beach. A more specific description of frequency in these contexts would be temporal frequency – the rate of repetition in time.

Figure 4.1 illustrates how we can often simplify the description of a complex *time domain* signal (e.g. a recorded sound wave) by representing it as a *spectrum* showing the relative intensities of its individual frequency components. In this figure we see just one time dimension – the *x* axis. Although less familiar in everyday life, 2D examples are common in science. In magnetic resonance spectroscopy for example, it is normal to add third and even higher order time dimensions to investigate molecular structures.

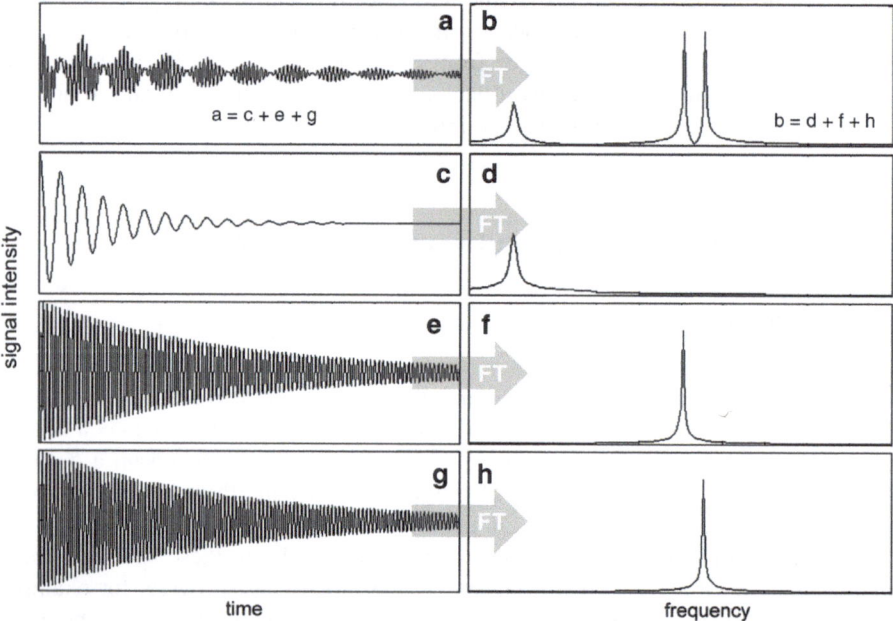

**Fig. 4.1** Signals that have a regular periodic variation of intensity over time (*left*) are often represented as *spectra* (*right*) that show the intensities of specific frequency components. The most common method of conversion of a time domain signal to its frequency domain representation is the *Fourier transform (FT)*. In this example the rather complicated *time domain* signal (**a**) has a quite simple *frequency domain* representation (spectrum **b**) containing just three distinct frequency components. These three components, shown singly in spectra **d**, **f**, and **h**, correspond to the time domain signals **c**, **e**, and **g** respectively, and each of these is a simple decaying sinusoid. Signal **a** is the sum of the signals **c**, **e**, and **g**. Physical phenomena that might produce these signals could be as diverse as the vibrations of a musical instrument, or nuclear magnetic resonance in a solution of small molecules

Going beyond the familiarity of the idea of temporal frequency there is no reason we can't apply the same concepts (and the same maths) to signals that change with space. Our brains can do quite a good job of resolving sounds into their temporal frequency components, i.e. recognizing notes. Some people can do this with extreme precision. Human brains have, however, not evolved to perform spatial frequency analysis of what our eyes perceive. For that we need an external tool – preferably a computer.

### 4.2.3  The Concept of Spatial Frequency

In image processing we often use the term frequency to describe the rate of change of a signal in space, for example the rate at which the pixel intensity changes as we scan across or down an image. In this context we are talking about *spatial frequency* (in fact if we scanned the image and recorded the change of intensity then we would once again have a signal that changed with time and thus the spatial and temporal frequencies would be directly related).

The concept of spatial frequency is extremely useful in image processing. Many of the methods used in analog and digital signal processing (signals often described by their temporal frequency) have direct equivalents in image processing. This transfer of methods from the temporal frequency domain to the spatial frequency domain means much of the terminology has come along for the ride, with occasionally confusing consequences. We will encounter high-pass, low-pass, band-pass and ideal filters in image processing just as we would in (temporal) signal processing.

Let's start with a very simple spatial frequency example. Consider the image shown in Fig. 4.2. As we scan the image from left to right the intensity starts initially at mid gray, increases slowly to white, decreases slowly to black, and then increases again to mid gray. If we were to plot the intensity across one row of the image matrix (all the rows are identical in this image) we would see that the profile has a sinusoidal shape. It looks very similar to the trace we would see if we connected an oscilloscope to the domestic electricity supply and measured the changing voltage. We can say that the intensity changes with a particular spatial frequency – in this case the frequency is one cycle per image width. We can describe this particular intensity modulation with a very simple mathematical expression *of the form*:

$$I = sin(\omega x) \qquad (4.1)$$

where $I$ is the intensity,
$x$ is the distance across the image,
and $\omega$ is the spatial frequency.

Since the value of $sin(\omega x)$ ranges from $-1$ to $+1$, in Eq. 4.2 an intensity of $+1$ would represent white, and $-1$ would represent black.

If the bit depth of our image data was 8 we would have $2^8 = 256$ possible intensities and we would normally use the convention $0 = $ black and $255 = $ white

**Fig. 4.2** This is a very unusual image because the intensity profile in the $x$ direction (a plot of the pixel intensities in any one row of the image matrix) is a perfect sinusoid with wavelength identical to the width of the image. The pixel intensities ($I$) can be described with a simple expression of the form: $I = sin(\omega x)$ where $\omega$ is the *spatial frequency* of the intensity variation. In this particular image $\omega = exactly$ one cycle per image width

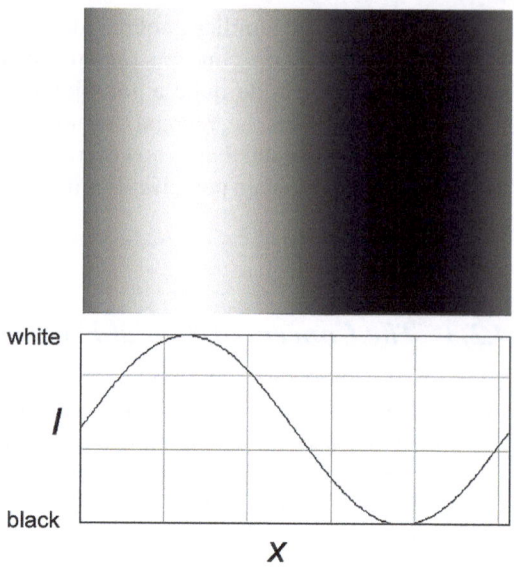

when displaying the image. In this case the precise expression for the x-direction change in intensity in Fig. 4.2 would be:

$$I = 127.5 \times sin\left(2\pi\omega\frac{x}{m}\right) + 127.5 \tag{4.2}$$

where $x$ is the distance across the image in pixels,
$m$ is the width of the image in pixels,
and the factor $2\pi$ is introduced to convert the spatial frequency from cycles per image width to radians per image width.

To keep this introduction as simple as possible we will just say that the maximum of the expression represents black and the minimum white. Also we wont worry about the need to convert cycles to radians so can leave out the $\frac{2\pi}{m}$. We can ignore these details for now as we are mainly interested in the general form of the intensity change as we move through the space of the image.

Now let's look at the slightly more complex image shown in Fig. 4.3. This image is similar to Fig. 4.2 but we now see small modulations in the intensity superimposed on the single cycle that spans the image. Now the intensity is modulated with two different spatial frequencies – a low frequency of 1 cycle per image width, and a higher frequency of 10 cycles per image width. Note that the amplitude of the higher frequency modulation is about one fifth of that of the low frequency modulation. In this case the (simplified) expression for the intensity modulation would have two terms, one for each frequency:

$$I = sin(\omega_1 x) + \frac{1}{5}sin(\omega_2 x) \tag{4.3}$$

**Fig. 4.3** A modified version of Fig. 4.2. The original intensity profile has a superimposed 'ripple' that can also be described by a sinusoid. The ripple pattern has *higher spatial frequency* and *lower amplitude* than the underlying profile. In this image the intensity profile can be described as the sum of the two sinusoids of different spatial frequency and amplitude:
$I = sin(\omega_1 x) + \frac{1}{5} sin(\omega_2 x)$,
where $\omega_2 = 10\omega_1$

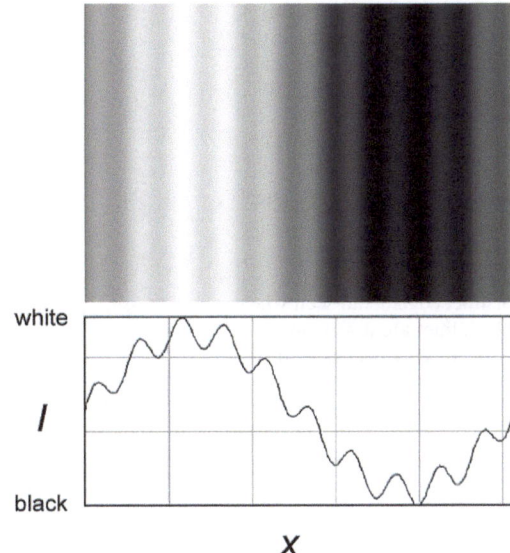

white

$I$

black

$x$

where $\omega_2 = 10\omega_1 = 10$ cycles per image width

Once again $\omega_1$ has the value of one cycle per image width, or $\frac{1}{m}$ cycles/pixel. The astute will notice that this expression has a maximum value of about 1.2 – whiter than white in our $1 = white, -1 = black$ scheme. We can ignore this for now as we are mainly interested in the general form of the intensity change as we move through the space of the image.

Look closely at the intensity profile in Fig. 4.3. The addition of the higher spatial frequency (10 cycles per image width) means that the intensity changes much more quickly in the $x$ direction than it does in Fig. 4.2 – the higher spatial frequency represents more rapid changes in image intensity as we scan a row of the image matrix. Put another way, the maximum *steepness* of the intensity gradient increases with the spatial frequency.

If we add a still higher spatial frequency, as in Fig. 4.4, we obtain a pattern that re-sembles a series of narrow black and white lines. As the spatial frequency increased so too did the maximum rate of change of image intensity.

Because the intensity modulations in the above images have sinusoidal profiles the mathematical expressions for the modulation are particularly simple. In fact the choice of sinusoidal intensity modulations in these introductory images is very de-liberate. They illustrate the idea of describing changes of intensity in space with a sine wave. We could also make up simple images in which the intensity change was most simply described as the cosine of a spatial frequency and the position in space. A more complex image might be described by a mixture of sine and cosine terms.

This leads us to a **VERY** important principle of image processing. *No matter what the intensity profile of an image might be it is possible to describe it as the sum of a collection of sine and/or cosine waves of different frequencies and amplitudes.* Before we introduce this idea more formally, let's consider how a collection of sinusoids can describe an intensity profile that seems very different from a sinusoid – Fig. 4.5.

**Fig. 4.4** The profile of this image can be described as the sum of two sinusoids of spatial frequencies 1 and 20 cycles per image width: $I = sin(\omega x) + sin(20\omega x)$. In this example both the spatial frequencies have equal amplitude. Notice that the higher spatial frequency component resembles the distinct edge detail seen when black lines are drawn on a white or gray background

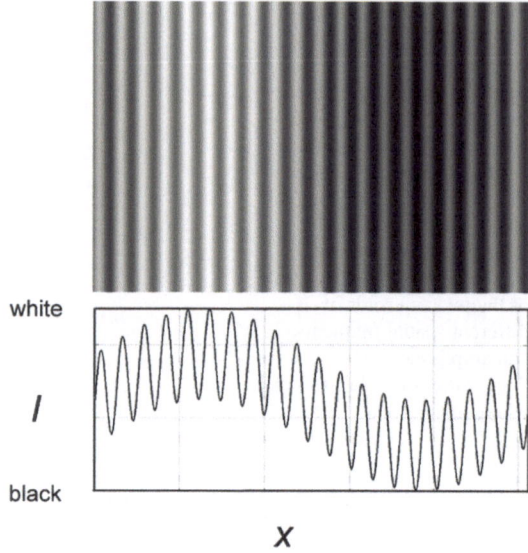

**Fig. 4.5** The intensity profile of this image does not resemble a sinusoid, yet, as the following images demonstrate, it can be accurately described as the sum of a long series of sinusoids of increasing spatial frequency and decreasing amplitude

Here the intensity profile has a step-like change from white to black exactly midway across the image. Although the profile looks nothing like a sinusoid we will soon see that it can be represented as the sum of a large number of sinusoids of progressively increasing frequency.

The first step in the process is to find a *single* low frequency sinusoid that gives a rough approximation to the step shaped profile. For this particular example the sinusoidal profile of Fig. 4.2 is just what we are looking for – it is bright or white in its left half and dark or black in its right half – not a very accurate representation of Fig. 4.5, but not too bad either. How can we improve it? Looking at the sinusoidal profile we could say that we need to flatten out the top of the white hump and the bottom of the black trough. We can do this by adding another sinusoid of *exactly* three times the frequency of the original as shown in Fig. 4.6. To achieve the optimum flattening of the white hump and the black trough the higher spatial frequency component has to have one third of the amplitude of the original. We still

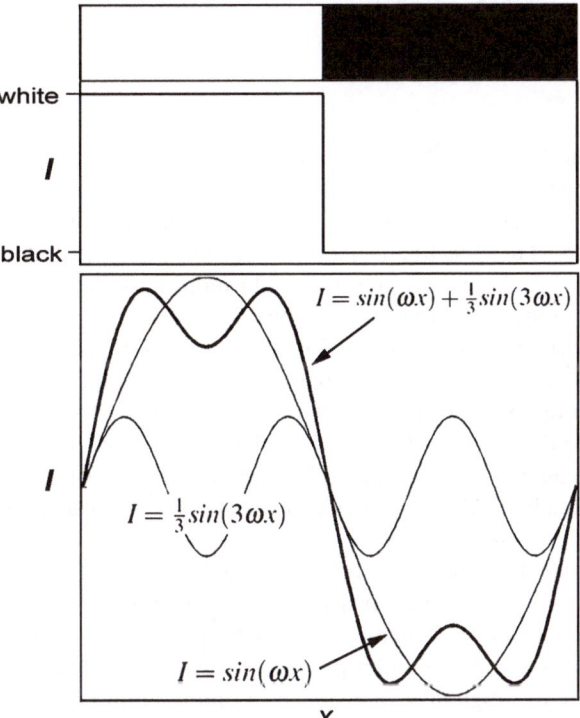

**Fig. 4.6** The square intensity profile of Fig. 4.5 can be approximated by adding two sinusoids of frequency $\omega$ and $3\omega$, where $\omega = 1$ cycle per image width. Addition of the higher frequency term flattens the top and bottom of the profile and steepens the sides. The best approximation to the square profile is obtained when the amplitude of the higher frequency is $\frac{1}{3}$ of the amplitude of the lower frequency

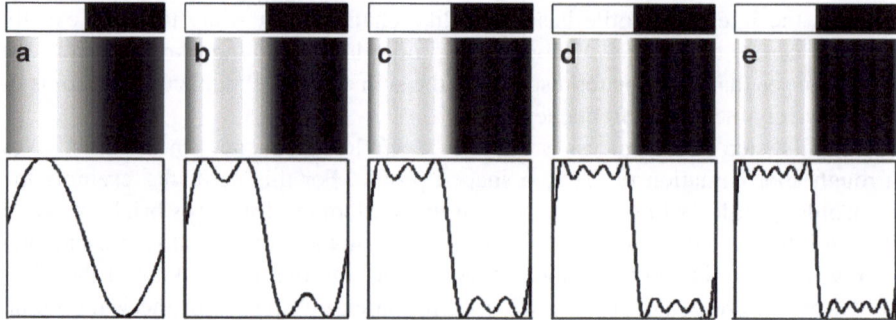

**Fig. 4.7** Five different sinusoid-based profiles that approximate Fig. 4.5. As each successively higher spatial frequency term is added the accuracy of the approximation improves

a   $I = sin(\omega x)$
b   $I = sin(\omega x) + \frac{1}{3}sin(3\omega x)$
c   $I = sin(\omega x) + \frac{1}{3}sin(3\omega x) + \frac{1}{5}sin(5\omega x)$
d   $I = sin(\omega x) + \frac{1}{3}sin(3\omega x) + \frac{1}{5}sin(5\omega x) + \frac{1}{7}sin(7\omega x)$
e   $I = sin(\omega x) + \frac{1}{3}sin(3\omega x) + \frac{1}{5}sin(5\omega x) + \frac{1}{7}sin(7\omega x) + \frac{1}{9}sin(9\omega x)$

have humps and troughs but their depth, or magnitude, is smaller than in our first, single sinusoid approximation to Fig. 4.5.

Once again we can improve the fit by flattening humps and troughs with higher frequency terms. The first five steps of this process are illustrated in images and profiles in Fig. 4.7. As we progressively add more terms of higher frequency we achieve an ever closer approximation to the step-shaped profile of Fig. 4.5. In *this particular* case (two bands of identical width) each new higher frequency term is an odd multiple of the first single-cycle approximation, and each new term has a smaller amplitude than the previous one. As we add each new term the 'roundness' of the intensity profile decreases and so too does the amplitude of the ripples. Notice that as each odd frequency term is added the spatial position of the mid gray intensity level remains unchanged and coincides exactly with the black/white step and the edges of the original image.

In this process the choices of frequency and amplitude are not arbitrary – they are each chosen to give the best possible approximation to the original step shape. In this particular example image the spatial frequency terms are *odd harmonics* - exactly analogous to the harmonics observed in a vibrating string.

Of course the rectangular 'square wave' profile seen in Fig. 4.5 would be highly unusual in a medical image. To describe intensity profiles seen in conventional images we would need more frequencies than just the odd harmonics. In fact, even Fig. 4.5 is a special case because the white to black transition occurs precisely midway across the image. If it did not then the odd harmonics alone would not be sufficient for an accurate representation – as we shall see later. (Note: Don't confuse the frequencies and amplitudes we used to represent of Fig. 4.5 with those that describe Figs. 4.3 and 4.4 which have quite arbitrary spatial frequency components.)

We have just examined some simple images and seen how we could describe their intensity profiles as a single sinusoid or the sum of a series of sinusoids of particular amplitude and spatial frequency. We only looked at the profile in the $x$ direction as these images had no variations in intensity in the $y$ direction. Most images aren't so simple – the intensity is likely to be highly variable in both the $x$ and $y$ directions, so for a proper description of an image's spatial frequency components we need a 2D method. There are several to choose from. In this text we will concentrate on the Fourier Transform because of its versatility and wide usage in image construction and processing. However, the 2D Fourier Transform will be a little easier to relate to the discussion we've just had if we first glance over the Cosine and Hartley Transforms. Think of it as learning to wrestle a salt water crocodile by first playing with a couple of goannas.

### 4.2.4 The Cosine and Hartley Transforms

The *Cosine Transform* is a rigorous method for doing what we have just done by intuition with the main difference being that it describes a profile as the sum of a series of cosine terms. Just for completeness, and not for memorization, here is the mathematical description of the discrete 1D Cosine transform:

$$F(\omega) = \sqrt{\frac{2}{N}} \sum_{x=0}^{N-1} f(x).c_\omega \cos\left(\pi \frac{\omega(2x+1)}{2N}\right) \tag{4.4}$$

This equation says that if we have a digital image matrix $N$ pixels wide then we can accurately describe the $x$ profile of any row of the image matrix by adding together $N$ different spatial frequencies. An analogous *Sine Transform* also exists, and although it more obviously resembles the intuitive process we followed in the last section, it is rarely used in image processing because it has mathematical disadvantages that make the Cosine Transform more useful.

Since we are dealing with digital images which are 2D arrays of discrete intensities the relevant implementation is the 2D *Discrete Cosine Transform* or DCT. The DCT is widely used in image processing – most commonly as a central part of the JPEG image compression method in which it is used to identify low amplitude high spatial frequency components that can be discarded because they have low visibility to humans.

In contrast to the Cosine Transform the *Hartley Transform* describes an intensity profile using both sine and cosine terms:

$$F(\omega) = \sum_{n=0}^{N-1} x_n \left[\cos\left(\frac{2\pi n\omega}{N}\right) + \sin\left(\frac{2\pi n\omega}{N}\right)\right] \tag{4.5}$$

Using both sine and cosine terms may seem like an unnecessary complication but we introduce it here because of the close relationship between the Hartley Transform and the Fourier Transform. In ImageJ (the image processing software used to illustrate examples in this text) the Fourier Transform is computed from a Hartley Transform.

## 4.3 Fourier Transforms and Fourier Spectra

With the preceding introductory background we can now discuss the use of the Fourier transform (FT) to resolve the spatial frequency components of 2D images.

### 4.3.1 1D Fourier Transforms

Put simply the Fourier transform states that any periodic function can be expressed as the sum of an infinite series of sines and cosines:

$$F(\omega) = \int_{-\infty}^{\infty} f(x)\,(cos(2\pi\omega x) - isin(2\pi\omega x))\,dx \qquad (4.6)$$

This, just one of several possible expressions for a 1D Fourier Transform, is included here mainly for the sake of completeness. You don't have to remember it, nor understand it in detail, to understand its use in image processing. Since an image is two dimensional we need to apply Fourier transforms in both directions, a complication that will be described soon. For now we need only consider 1D transforms.

A major difference between the Fourier Transform and the Hartley Transform is that the FT is *complex* – it includes the *imaginary* number $i = \sqrt{-1}$. If you are not familiar with the idea of complex numbers don't panic. For the curious there is a basic introduction to complex numbers included in Appendix C, but you don't need to know this to get a feel for what the Fourier transform is doing.

Why have we now introduced yet another headache – complex numbers? What is the point of imaginary numbers? After all, images are composed of real numbers – the intensities of pixels.

If we were *only* interested in processing images to determine their spatial frequency components then the Cosine or Hartley Transforms would indeed be sufficient. However, image processing and image construction needs more versatility than this. Although you might argue that imaginary numbers don't exist (you can't count or measure objects with them), the complex number formalism is indispensable. We need complex numbers and Fourier Transforms to make MRI images and they are very handy for image analysis. Take a look at Appendix C for some more examples.

The following discussion deals mainly with the *magnitude* or *absolute value* of the Fourier Transform. The FT of a purely real function (one that does not involve multiples of i) such as a digital image, has complex terms, but we need not concern ourselves with this technicality just yet.

Working in the opposite direction to the Fourier transform above, we can synthesize an arbitrary periodic function by *Inverse Fourier Transformation* of its frequency components:

$$F(x) = \int_{-\infty}^{\infty} f(\omega)\left(cos(2\pi\omega x) + isin(2\pi\omega x)\right) d\omega \qquad (4.7)$$

This is very similar to the forward or direct Fourier Transform, the difference being only the sign of the second term.

Equations 4.6 and 4.7 are descriptions that apply to *continuous* functions and are unsuitable for digital data which is *discrete*, not continuous, and *finite*, not infinite. To deal with discrete finite data sets such as digital images we use a modified version of the Fourier transform – the *Discrete Fourier Transform* (DFT):

$$F(\omega) = \sum_{x=0}^{N-1} f(x)\left(cos\left(\frac{2\pi\omega x}{N}\right) - isin\left(\frac{2\pi\omega x}{N}\right)\right) \qquad (4.8)$$

In this expression $x$ represents the pixel position where there are $N$ pixels in one row or column of the image, and $\omega$ represents a specific spatial frequency. Notice that in the discrete Fourier Transform the integration from $-\infty$ to $+\infty$ has been replaced by a summation from 0 to $N - 1$ because we only have a finite set of data to deal with.

There is also a discrete form of the inverse Fourier Transform:

$$F(x) = \frac{1}{N}\sum_{\omega=0}^{N-1} f(\omega)\left(cos\left(\frac{2\pi\omega x}{N}\right) + isin\left(\frac{2\pi\omega x}{N}\right)\right) \qquad (4.9)$$

Using the DFT and inverse DFT we can convert an image back and forth between the spatial and frequency domains as many times as we like *without any loss of information*. The transforms are precise descriptions, not approximations. The ability to use such transforms without fear of data loss permits the processing of images in either the spatial or frequency domains. The only limitation to this claim is the internal precision of a computer's calculations. These are more than adequate for most image processing tasks.

The summations expressed in the DFT formulae can easily be performed by a computer. The DFT is still a complex function, and you might wonder how the computer represents $\sqrt{-1}$. It doesn't have to. We only need to store and manipulate the coefficients of the imaginary terms and those coefficients are real numbers. This, believe it or not, is how an anatomical image is constructed from raw MRI data in which the real and imaginary coefficients represent voltages measured on different axes of the MRI scanner.

The DFT has a special characteristic which is very important in image processing. *The DFT treats its finite set of discrete input data as if the data repeated itself infinitely.* For a digital image this means the image data is treated as if the image were tiled infinitely in space – a concept that will be illustrated shortly.

### 4.3.2   2D Fourier Transforms

For digital images (2D matrices of real numbers) we use a 2D Fourier transform to resolve the $x$ and $y$ spatial frequency components. The 2D FT produces a two layer matrix – one layer representing the coefficients of the real terms, and the other layer the coefficients of the imaginary terms.

Let's consider a digital image – an $m$ column $\times$ $n$ row 2D matrix of pixel intensities. To perform the 2D Fourier transform of the image we can first compute the 1D DFT of each row of the image matrix and store these transformed rows in a new two layer complex matrix representing the real and imaginary coefficients. This new matrix now has both space and spatial frequency dimensions. Each row represents the series of spatial frequencies present in the corresponding row of the original image. Each column represents the coefficients of a particular $x$ direction spatial frequency, and these may vary according to the $y$ coordinate (row number) in the original image. Since the DFT is a summation of a finite number of terms ($N$ in Eq. 4.8), not an infinite series, the DFT of a row of $m$ pixel intensities has $m$ terms representing $m$ distinct spatial frequencies.

To complete the 2D Fourier transform we now compute the 1D DFT of each of the $m$ columns of the complex matrix of row transforms. The output of this process is another complex matrix of spatial frequency coefficients. This matrix has spatial frequency dimensions in *both* directions – $x$ and $y$. The elements of any row or column of this matrix represent coefficients of spatial frequency in the original image *but they no longer have any direct correspondence to a particular row or column in the original image*. We need the inverse Fourier Transform to decode this information.

We could also have first FT'ed the columns of the image matrix, and then the rows of the matrix of column FTs, with the same final result. The actual algorithm used in most software differs a little from this rows-then-columns method for the sake of computational efficiency. It has the same effect except that the complex FT matrix is square and has dimensions $m' \times m'$ where $m'$ is a power of two – the $m' \times m'$ square is simply the smallest one that can enclose the original image.

### 4.3.3   Fourier Spectra

In Fig. 4.7 we partially resolved an image into its spatial frequency components (in one dimension). We stopped our approximation when we got to five terms, but adding more higher frequency terms would have increased the accuracy of the approximation. We could have used a 2D Fourier transform to perform the same task, including calculation of all the higher order terms, with a result as represented in Fig. 4.8. The FT is a complex function and often produces negative coefficients for the real and imaginary terms. To illustrate these negative coefficients in Fig. 4.7 the real and imaginary parts of the complex data are represented by two special images (c, d) in which zero is represented by mid-gray rather than black. Positive coefficients are represented by lighter pixels and negative coefficients by darker pixels. While we have illustrated the real and imaginary data separately in this figure, they are not normally viewed in this way. The convention is to show the *magnitude* or *Fourier spectrum* which represents the square root of the sum of the squares of the real and imaginary coefficients of the complex matrix (for more explanation of this see Appendix C). The magnitude is thus always a positive number (we squared the coefficients, not the imaginary numbers) and is thus easy to display. This is what is shown in Fig. 4.8b. Take note of the *spectrum* part of the description 'Fourier spectrum'. A spectrum is a display of intensity as a function of frequency – in our case,

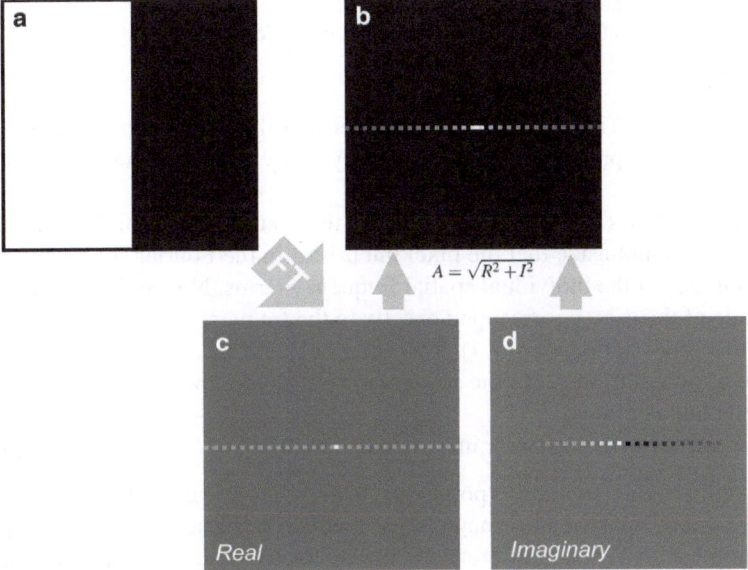

**Fig. 4.8** 2D Fourier transform of Fig. 4.5. The Fourier spectrum (**b**) represents the *magnitude* of the Real and Imaginary parts, represented here by images **c** and **d**. The Fourier spectrum amplitude $A = \sqrt{R^2 + I^2}$ where $R$ and $I$ are the coefficients of the real and imaginary terms. Note that in **c** and **d** mid-gray represents zero. Positive coefficients are represented by lighter pixels and negative coefficients by darker pixels. For *this particular* image all the real coefficients are positive. The Fourier spectrum of a purely real matrix, such as a digital image, is always symmetrical – even if the image is not symmetrical. Detail of the center of this Fourier spectrum is shown in Fig. 4.9

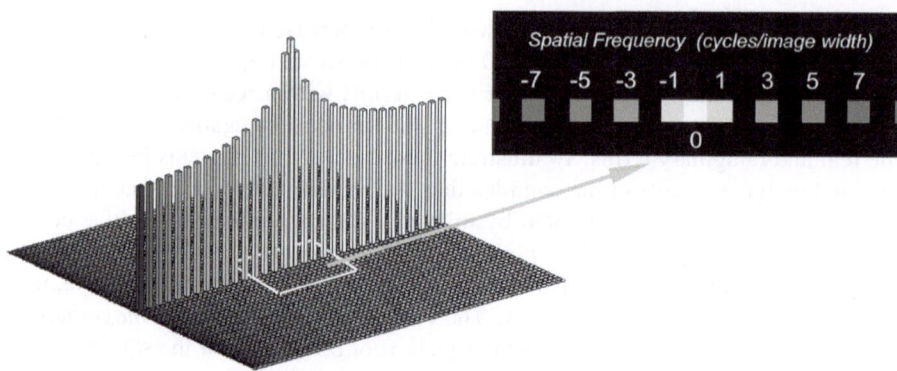

**Fig. 4.9** An alternative representation of Fig. 4.8b. On the *left* the intensity of the pixels is plotted as a 3D bar chart. In a Fourier spectrum it is conventional to plot the log of the complex amplitude so that the normally large zero and low frequency amplitudes does not obscure the display of the terms with smaller amplitudes. On the *right* we see an enlargement of the center of the Fourier spectrum image. Note that the intensities of the pixels representing amplitudes of the spatial frequencies 1, 3, 5, 7, etc. *decrease* as frequency increases – just as we saw in Fig. 4.7 (the height of the bars, and corresponding pixel intensities, decrease slowly because it is the log of the amplitude that is plotted). The bright zero frequency point at the center of the Fourier spectrum represents (but is not equal to) the average of all pixel intensities in the original image. Because it has the largest amplitude it is always shown as a white pixel in the Fourier spectrum

spatial frequency. This should be a reminder that the $x$ and $y$ axes of the Fourier spectrum image represent $x$ spatial frequency and $y$ spatial frequency, NOT spatial distances.

Figure 4.9 shows two alternative views of the Fourier spectrum. The 3D bar chart is intended to emphasize that the pixel intensities in the Fourier spectrum represent the amplitudes of the individual spatial frequency terms. Notice that the first five on either side of the center correspond exactly to the frequencies and amplitudes of our intuitive approximation (Fig. 4.7). The height of the bars, and corresponding pixel intensities, do not drop off in the sequence 1, $\frac{1}{3}$, $\frac{1}{5}$, $\frac{1}{7}$, ... because it is the log of the amplitude that is plotted.

There are some important points to note about the Fourier spectrum display:

- The origin ('zero frequency point', see below) lies by convention at the center of the FT image matrix. The image is symmetrical because in the Fourier spectrum of an image there is no difference between positive and negative frequencies.
- *For this example* all the non-zero data lies along a central row at right angles to the black/white edge in the original spatial domain image.
- *For this example* the non-zero data appears in the first and then every second pixel as we count outwards from the central pixel. These are the odd harmonics mentioned previously.
- The pixel intensity decreases with increasing distance from the center. This represents the progressively decreasing amplitude of the higher order harmonics, also seen in Fig. 4.7.

- *For this example* there are no non-zero elements in the vertical direction apart from the central row. This is because there is no modulation of the intensity in the vertical direction, i.e. for any column of the image matrix for Fig. 4.5 all the elements are identical.

In general, if the dimensions of a digital image are $n \times n$ pixels, then there will be $n$ terms (i.e. $n$ discrete spatial frequencies) in each dimension of its Fourier transform. Many of these spatial frequencies may have zero amplitude. The equations in the caption of Fig. 4.7 only show the non-zero terms.

### 4.3.4   The Zero Frequency or 'DC' Term

What about the central white pixel we see in all the Fourier spectra? As mentioned above, the center of the FT matrix represents the 'zero frequency' and is often called a DC term because of its *direct current* (DC) equivalent in electrical signal processing. It represents a signal that does not vary with time or space. In the Fourier transform of an image the DC term can be thought of as representing the average intensity value of the whole image. The only image that will have a zero amplitude for the zero frequency is a completely black image.

The real explanation for the zero frequency term is hidden in Eq. 4.2. Remember that the value of sines and cosines range from $-1$ to $+1$ but in real images the pixel intensities are all positive numbers or zero. To keep our mathematical description of image profiles simple we temporarily adopted the convention of $-1 =$ black and $+1 =$ white. Going back to the $0 =$ black and $255 =$ white convention for an 8 bit image we need to add a constant to the sum of all the sine terms to make sure the total is non-negative. In Eq. 4.2 we did this by adding 127.5 to the sine term that describes the profile of Fig. 4.2. Notice that 127.5 is the average pixel intensity. Thus the DC term represents a signal that does not vary with space and it's effect is to 'offset' the output of the non-zero frequency terms.

Because the amplitude of the zero frequency term is usually very much higher than the amplitude of any of the other frequency terms it is common to plot the log of the amplitudes to display an image of the Fourier spectrum. This makes it easier to see variations in the amplitude data.

### 4.3.5   Fourier Spectra of More Complex Images

In Fig. 4.10 we see a slightly more complicated version of Fig. 4.5 and its Fourier spectrum. At first sight this Fourier spectrum looks identical to the one shown in Fig. 4.10b, but if we look closely we see that in Fig. 4.10b the pixels immediately on either side of the central (zero frequency) pixel are black. There is a good reason for this. The first non-zero term in the sinusoidal approximation to the intensity profile of Fig. 4.10a has twice the frequency of that for Fig. 4.5 because in Fig. 4.10a

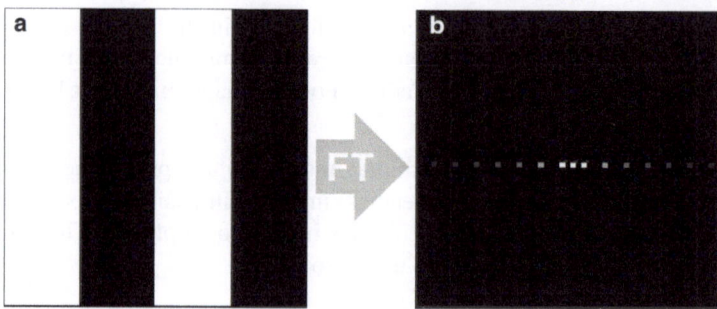

**Fig. 4.10** Another simple image and its Fourier spectrum. For this image the spatial frequency ±1 cycle per image width has amplitude zero – it is not useful in describing the *x*-direction profile. This profile is best described by summation of spatial frequencies ±2, ±6, ±10, ±14, etc

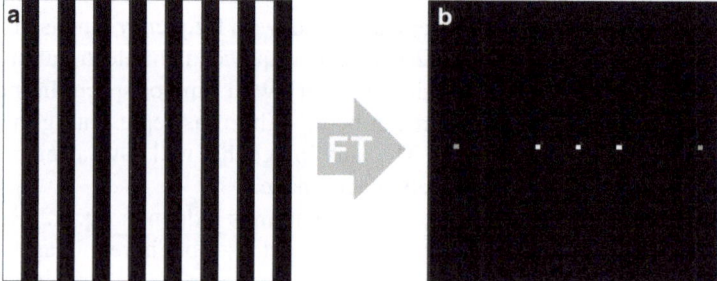

**Fig. 4.11** A more complex image and its Fourier spectrum. For this image the spatial frequencies from ±1 to ±7 cycles per image width have amplitude zero – they are not useful in describing the *x*-direction profile. This profile is best described by summation of spatial frequencies ±8, ±24, ±40, etc. For clarity image **b** shows an enlargement of the central part of the Fourier spectrum

we have two black and two white bands. The amplitude of the spatial frequency 1 cycle/image width is zero. If we increase the number of black and white bands further (Fig. 4.11) the first non-zero terms in the FT occur, as we now expect, further from the center. We still have a 'square wave' intensity profile so the next non-zero frequency is the next odd harmonic of the 'primary' frequency.

Let's look again at why there is only one row of non-zero data in these Fourier spectra. Remember that to perform the 2D FT we first transformed all the *rows* of the original spatial domain image and then transformed all the *columns* of the intermediate row-FT matrix. Since all the rows of the original image are identical to each other, all the rows of the intermediate row-FT matrix will be identical to each other. Thus all the elements in any single *column* of the row-FT matrix will be identical to each other. When we transform these columns of the intermediate matrix we produce only DC values because all the elements in any single column are identical. The FT of a series of constant non-zero terms has just one non-zero term, the DC term. In the Fourier spectrum the *y* direction DC term lies midway

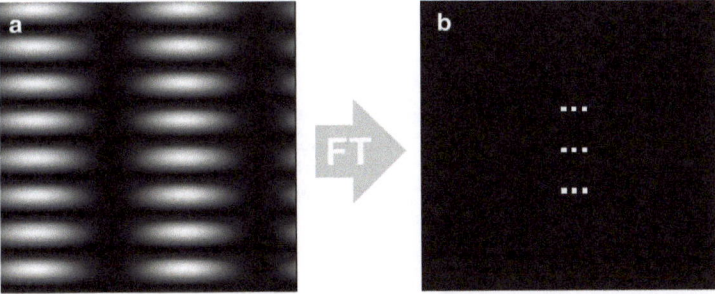

**Fig. 4.12** Most images have intensity changes in both the $x$ and $y$ directions. In this image we have a $y$-direction spatial frequency of exactly 8 cycles per image height, and an $x$ direction frequency of exactly 2 cycles per image width. For clarity image b shows an enlargement of the center of the Fourier spectrum. The size of the original spatial domain image matrix (**a**) was 128 × 128. If the full 128 × 128 Fourier spectrum were shown in **b** it would be difficult to see the details that are very close to the center

between the top and bottom. Of course this lack of spatial frequency modulation in the $y$ direction is also evident in the original image.

Now let's look at an image (Fig. 4.12) in which we have intensity changes in both the horizontal and vertical directions. Once again we have an intensity profile that is sinusoidal - two cycles per image width in the horizontal direction and 8 cycles per image width in the vertical direction. We see that the non-zero terms in the FT are displaced 2 pixels from the center horizontally and eight pixels vertically.

It should now seem reasonably obvious why we have the non-zero elements in the central row and central column of Fig. 4.12b, but what about the extra four white spots that have appeared off the center lines? We can understand the origin of these if we look at the 'halfway point' of the 2D Fourier transform process – the row-FT matrix. Figure 4.13 illustrates the stepwise creation of the Fourier spectrum of Fig. 4.12a. Since each of the Fig. 4.12a rows (except the ones that are completely black) has a sinusoidally modulated intensity with spatial frequency two cycles per image width the 1D FT has non-zero elements 2 pixels either side of the center. The completely black rows have 1D FTs that are all zeros. If we complete the 2D FT process by performing a 1D FT on every column of Fig. 4.13b (strictly speaking we mean on the complex matrix who's magnitude is Fig. 4.13b) we would find that, because only the central column and the two columns two pixels either side of the center have non-zero data, there will be only these same three columns with non-zero data in the final Fourier spectrum. Furthermore, this non-zero data in the intermediate matrix, as can be seen in Fig. 4.13b, has a sinusoidal profile with frequency eight cycles per image height. Thus in the final Fourier spectrum (d) we get non-zero elements eight pixels above and below of the center - exactly as we saw in Fig. 4.12b.

On the basis of the discussion so far we might assume that the Fourier spectrum of an image becomes more complicated as the complexity of intensity changes in

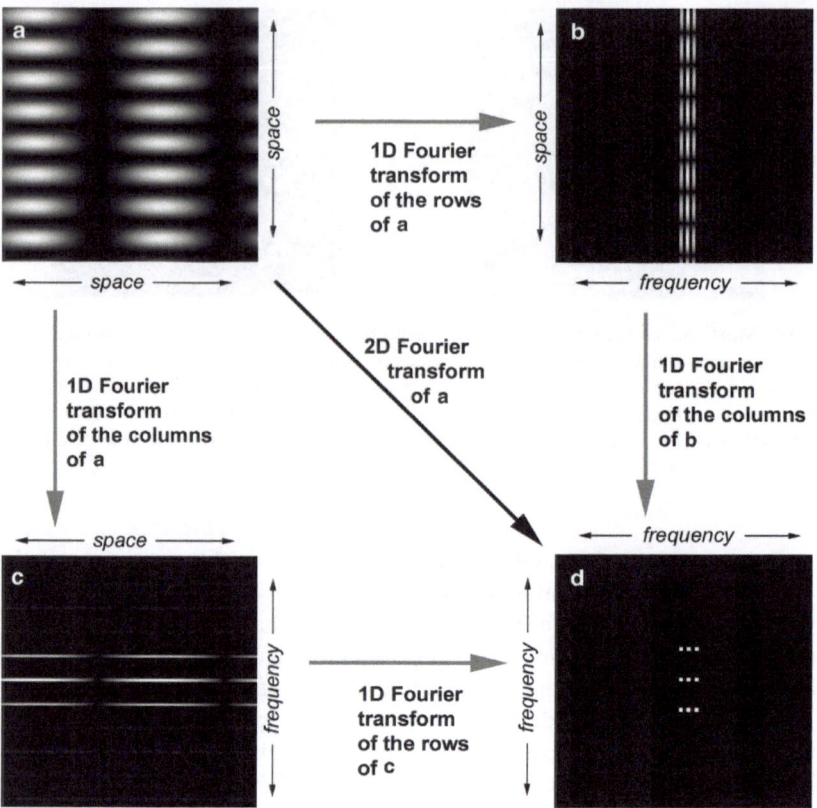

**Fig. 4.13** The 2D Fourier transform can be formed by stepwise 1D Fourier transformation. Either the rows or the columns can be transformed first to create intermediate 'partial Fourier spectra' (**b** and **c**). The coordinates of images **b** and **c** thus represent *spatial frequency* in one direction, and *space* in the other direction

the image increases, however, this is not always the case. Consider the image and Fourier spectrum shown in Fig. 4.14. Although this image appears quite simple in comparison with Fig. 4.12a its Fourier spectrum has so many non-zero terms that it looks like a white blur.

The explanation of this lies in our earlier statement that when we perform a 2D Fourier transform on an image the image is treated as if it were tiled infinitely in all directions. For the images we have discussed so far this tiling does not result in any change to the modulation of the intensity that we see within the image itself. Figure 4.14a is different from the previous images in this respect. Here we have a spatial frequency of 0.5 cycles per image width and 2.5 cycles per image height. Since the intensity modulation is not an integer number of full cycles in either direction the intensity profile of the infinite tiling does not have a regular sinusoidal profile. Figure 4.15 shows the tiling effect with just four copies of Fig. 4.14a. There are sharp discontinuities in the sinusoidal profile at the joins between tiles.

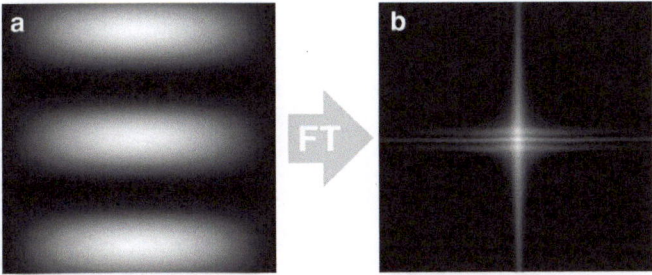

**Fig. 4.14** An apparently simple image may have a complicated Fourier spectrum containing non-zero amplitudes for a large number of spatial frequencies. This is because the Fourier transform treats an image as if it were tiled infinitely in all directions. Tiling of this particular image, as shown in Fig. 4.15 below, results in sharp discontinuities in the sinusoidal profile at the joins between tiles

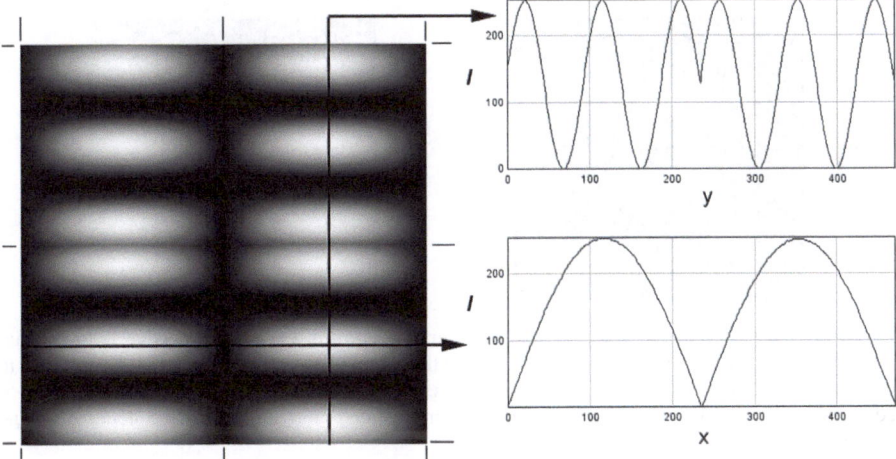

**Fig. 4.15** Partial tiling of Fig. 4.14a. The Fourier transform treats an image as if it were tiled infinitely in all directions. When Fig. 4.14a is tiled we find that the profiles in both the $x$ and $y$ directions feature sharp discontinuities where edges of the original image join. These discontinuities can only be described by the sum of a very large number of spatial frequencies, as evident in the Fourier spectrum (Fig. 4.14b)

The FTs of these 'corrupted' sinusoids contain a large number of non-zero amplitudes of many frequencies in order to account for the discontinuity in the intensity profile. The Fourier spectrum (Fig. 4.14b) looks like a white blur because many spatial frequencies have non-zero amplitudes.

Now consider Fig. 4.16a which looks exactly the same as Fig. 4.5 but has a very different Fourier spectrum – it appears to have no non-zero terms in the central row. This tells us something about Fig. 4.16a that we probably can't see with the naked eye. The black and white stripes are not exactly the same width (the black stripe in the original image was 33 pixels wide and the white stripe 31 pixels). Now we can no longer approximate the intensity profile with a series of odd harmonics because the black/white step does not occur precisely at a null (zero) point of the odd

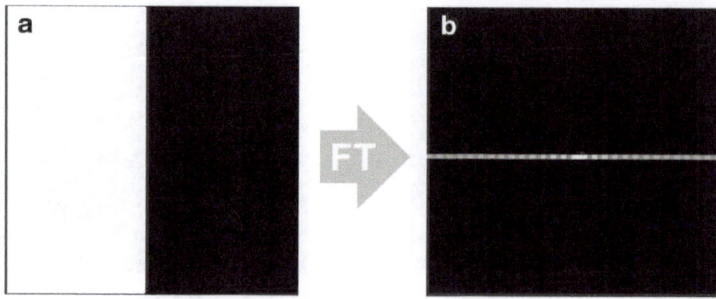

**Fig. 4.16** A slightly asymmetrical version of Fig. 4.5 and its Fourier spectrum. In this image the *white band* is slightly narrower than the *black band*. The Fourier spectrum shows non-zero amplitudes for *all* $x$-direction spatial frequencies. Unlike Fig. 4.5, this image cannot be described by the odd spatial frequencies alone

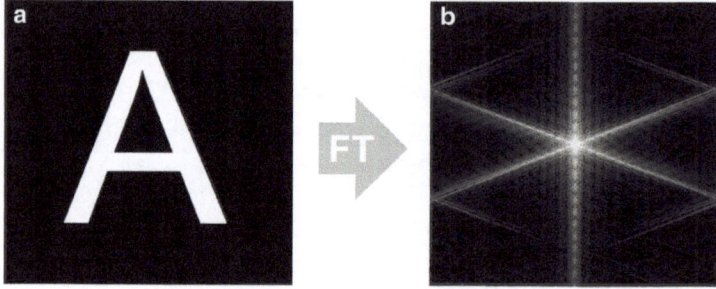

**Fig. 4.17** An image of the letter 'A' and its Fourier spectrum. Each straight edge in the spatial domain image gives rise to a linear feature (a long line of non-zero spatial frequency amplitudes) in the Fourier spectrum. The Fourier spectrum feature is always oriented perpendicular to the straight edge it describes in the original image. This is demonstrated more clearly for the 'letter A' image in Fig. 4.18

harmonics. In fact we need both sine *and* cosine profiles. This example demonstrates how examination of the Fourier spectrum can sometimes tell us things about the spatial domain image that are difficult or impossible to perceive directly.

Applying what we have learnt about interpretation of Fourier spectra we can now explain the features of Fig. 4.17. Because the edges of the letter are very sharp there are a large number of non-zero terms in the Fourier spectrum. Each of the straight lines that form the letter 'A' give rise to a blurred white line in the Fourier spectrum, and this line lies perpendicular to the edge from which it originated in the spatial domain image.

Just as the spatial domain image is treated by the 2D Fourier transform as if it were tiled infinitely in space the frequency domain data is also, effectively, infinitely tiled in frequency space. This means that the long diagonal features in the Fourier spectrum are 'wrapped' at the edges – a feature that extends to the edge of the Fourier spectrum continues, with decreasing intensity, from the opposite edge. This characteristic appears more obvious when we look at the Fourier spectra of the individual parts of the letter 'A' image (Fig. 4.18). Most importantly, note that the

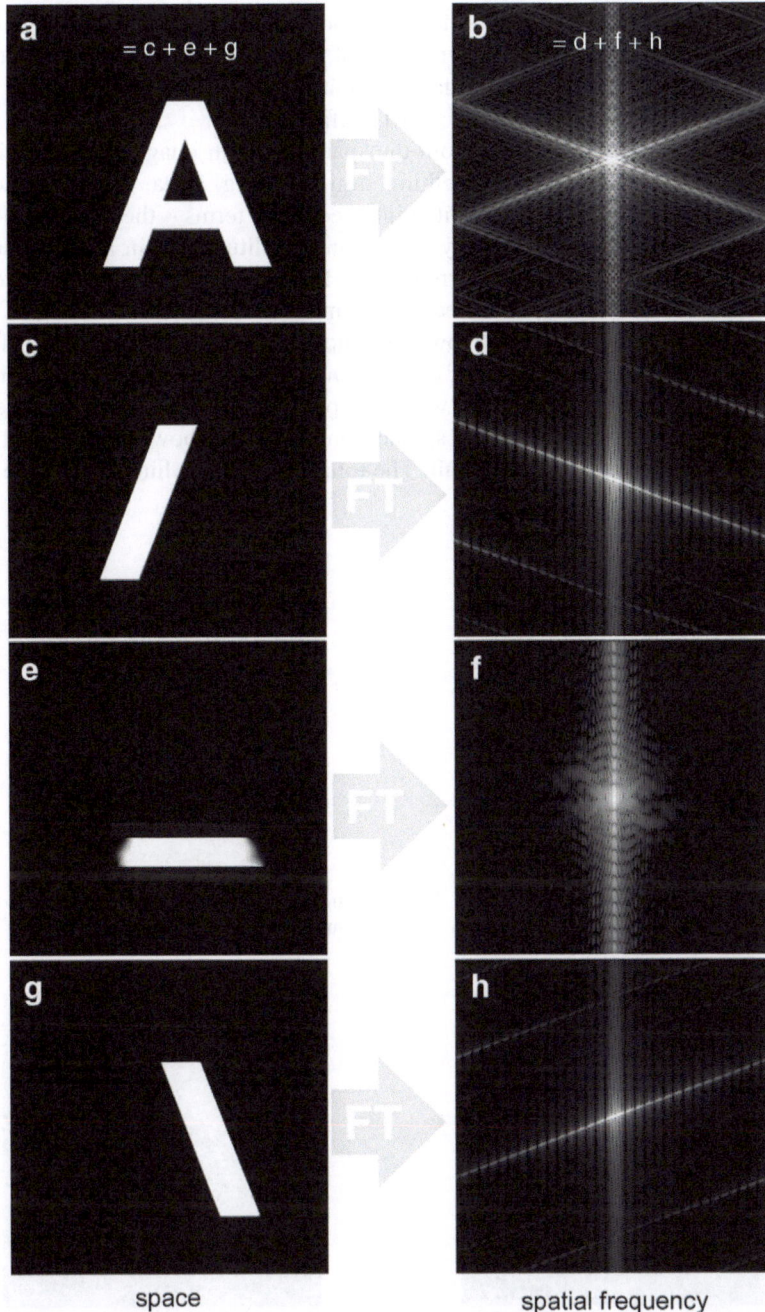

**Fig. 4.18** Fourier spectra of the individual parts of the letter 'A' image. Each long diagonal feature in the Fourier spectrum 'wraps' to the opposite side of the spectrum. Note that the mathematical processes represented in this diagram are essentially identical to those in Fig. 4.1. This diagram illustrates *2D spatial frequency* spectra and their corresponding *spatial* domain signals. Figure 4.1 illustrates *1D temporal frequency* spectra and their corresponding *time* domain signals

extended Fourier spectrum feature does not 'bounce' or 'reflect' at the edge of the Fourier spectrum. It is merely the symmetry of the letter 'A' image that leads to this illusion. As we will see later, this 'data wrapping' phenomenon is important in the formation and interpretation of some MRI artifacts.

If we remove the sharp edges from the spatial domain image by application of a blurring filter we might get something similar to Fig. 4.19a. Now we find that the Fourier spectrum loses most of its high frequency terms – their amplitudes are zero or very close to zero. The only significant amplitudes lie in a cluster around the central zero frequency. The blurred image has no rapid changes in its intensity profile – there are no high spatial frequencies in this image.

Conversely, we could zero the low frequency data from the FT of Fig. 4.17b as shown in Fig. 4.20a, where the black circle represents the area in which the complex FT data has been set to zero. Now when we perform an inverse FT on the edited frequency domain data we form the spatial domain image shown in Fig. 4.20b. The result is that we have only edge detail. The tonal detail, the white infill of the letter A, is lost.

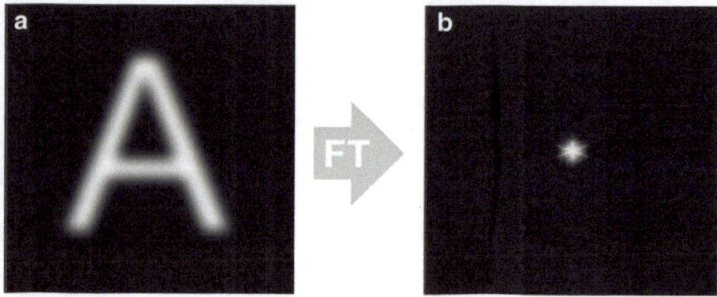

**Fig. 4.19** A blurred version of the 'letter A' image and its Fourier spectrum. There are no rapid changes in the intensity profile so only low spatial frequencies have non-zero amplitudes

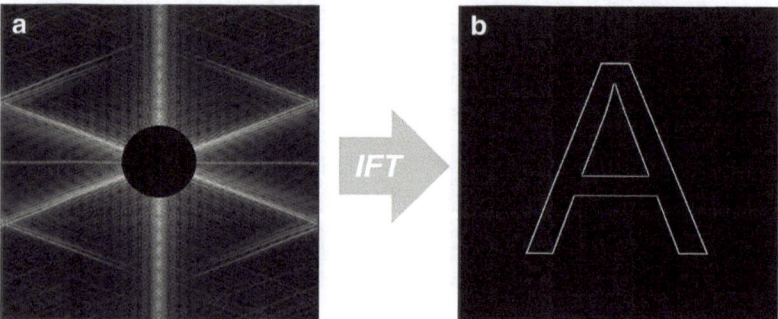

**Fig. 4.20** Edited FT data from Fig. 4.17b (**a**). All of the low frequency amplitudes under the black circle have been zeroed. Inverse Fourier transformation of this edited data produces a spatial domain image (**b**) that contains only edge information

### 4.3.6   How Many Spatial Frequencies are Needed?

2D Fourier spectra are images in which the pixel intensities represent the amplitudes of complex spatial frequency data. By convention the zero frequency point is plotted in the center of the Fourier spectrum. The pixels immediately adjacent to the central pixel represent the amplitudes, in the $x$ and $y$ directions, of the spatial frequency 1 cycle per image width. The next pixels represent the amplitudes of spatial frequency 2 cycles per image width, and so on. The amplitudes of the highest spatial frequencies are represented by the pixel intensities at the edges of the Fourier spectrum. What is the highest spatial frequency available? Is it high enough to describe the most sudden intensity change in the image?

Imagine that we have a square digital image which has dimensions $m \times m$ pixels. The Fourier spectrum of this image will have the same dimensions so the *maximum* spatial frequency will be $f_{max} = \frac{m-1}{2}$ cycles per image width (we have to subtract 1 because one pixel is assigned to the zero frequency 'DC' term). For a $65 \times 65$ pixel image this means the highest spatial frequency in the Fourier spectrum is 32 cycles per image width. Is this high enough?

Now imagine the highest possible rate of intensity change in a $65 \times 65$ pixel image. This would be when adjacent pixels in the image had the maximum possible intensity difference – such an image might look like Fig. 4.21, which is made up of alternating rows of black and white pixels. The total number of *line pairs* is 32.5. A *theoretical* way to describe this pattern with sinusoids would be to start with the spatial frequency 32.5 cycles per image width and add its odd harmonics, starting with $3 \times 32.5 = 97.5$. But we don't have *any* of these spatial frequencies available! Our highest is 32.

The solution is to use a sinusoid of frequency 32 and its *lower* frequency neighbors. This time the required amplitudes, as shown in Fig. 4.21, decrease as the spatial frequency *decreases*. In an $m \times m$ pixel image there can *never* be an intensity pattern denser than $\frac{m}{2}$ line pairs per image width, and this pattern can be described *perfectly* by addition of sinusoids of spatial frequency zero to $\frac{m-1}{2}$ cycles per image width.

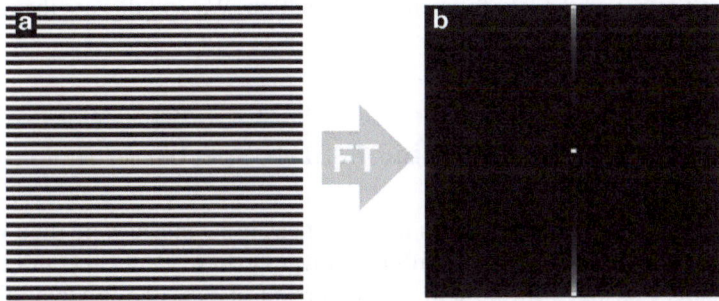

**Fig. 4.21** The maximum possible rate of intensity change in a $65 \times 65$ pixel image is 32.5 line pairs per image width (or, in this case, height). Only the spatial frequencies 0 to 32 cycles are available, but these are sufficient to perfectly describe the line pairs pattern as the sum of a series of sinusoids

Up to the limit of precision of the calculations, the 2D discrete Fourier transform is a 100% accurate description of a digital image.

### 4.3.7  Fourier Spectra of Lines

Based on the discussion so far we could reasonably come to the conclusion that the edge detail in an image is encoded in the outer high frequency part of the Fourier spectrum and the tonal detail, or shading, is encoded by the central low frequency part of the spectrum. What then would the Fourier spectrum of a narrow line look like? When we zeroed the low frequency amplitudes in Fig. 4.20 we were left with just the outline of the original bold 'A' shape – more or less a line drawing of the solid shape. It looks like we just need the medium and high frequencies to define a line drawing but the reality is a little more complicated.

Instead of going straight to the Fourier spectrum of a narrow line let's look at what happens to the Fourier spectrum of our original black and white stripe image (Fig. 4.8a) as we make the white stripe progressively narrower. The effect is illustrated, in three steps, in Fig. 4.22. As the white stripe gets narrower its Fourier spectrum continues to have significant amplitudes across the range of frequencies. The original very simple harmonic pattern seen in Fig. 4.8b becomes more complex and there are actually *more* non-zero low frequency amplitudes. When the white stripe is very narrow (effectively a line) *all* the spatial frequencies have large amplitudes. In the extreme case, when the line is infinitely narrow (in a digital image this means one pixel wide), the amplitudes of all spatial frequencies are identical – the profile of the Fourier spectrum is a flat horizontal line of constant intensity.

Our original idea that we only need high frequency terms to describe a line has turned out to be inaccurate. In fact we need *the full range* of available frequencies to describe a narrow line. The reason that we didn't need the low frequencies to show the outline of the letter A is that this is, in fact, not a very accurate outline. Close inspection of the original of Fig. 4.20b would show a series of feint gray lines running parallel to all of the white lines. These 'ringing' artifacts are due to the fact that we removed the low frequency information. We will discuss them more in Chapter 7.

## 4.4  The Complex Data Behind Fourier Spectra

Because the Fourier transform treats image data as if the image were tiled infinitely in space it is quite possible for two different images to have identical Fourier spectra. For example, Fig. 4.23a shows a modified version of Fig. 4.8a in which the black region has been moved from the right side to the center of the image. The Fourier spectra of the original and the modified image are identical. In this example we moved the black region but we didn't change its width – it remains the same as the

**Fig. 4.22** Progressive narrowing of the white stripe of Fig. 4.8a changes the simple harmonic pattern in the Fourier spectrum. When the white stripe is just one pixel wide *all* x direction spatial frequencies have identical amplitude. In the case, not shown here, of an image containing only a single non-zero pixel intensity (in other words a point) all x and y direction spatial frequencies have identical non-zero amplitude, and thus all the pixels in the Fourier spectrum have identical intensity

total width of the two white regions. So long as the black and white areas have the same width, no matter where we put the black stripe the infinitely tiled patterns will appear identical. The spatial frequency amplitudes needed to describe the patterns will be identical so the Fourier spectra will also be identical.

The information that describes the spatial *position* of the image data is encoded in the difference between the coefficients of the real and imaginary data. Comparing the representations of these coefficients in Fig. 4.8c & d with Fig. 4.23c & d we see that these parts are not identical.

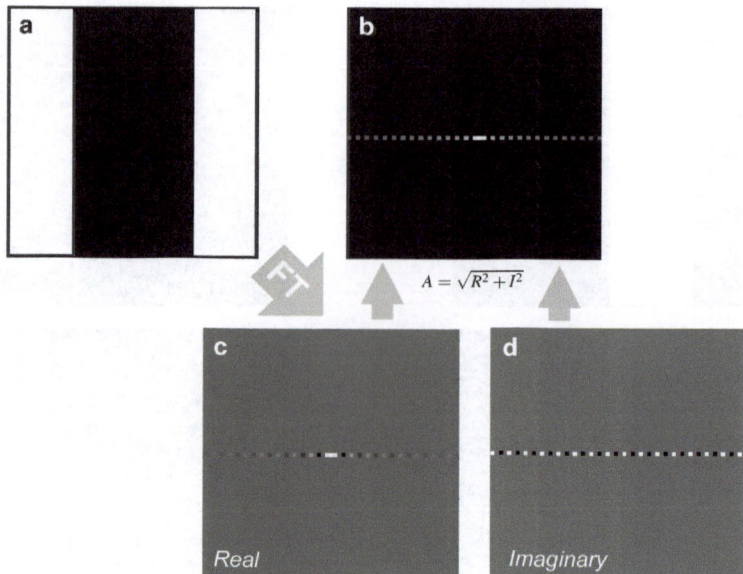

**Fig. 4.23** Shifting the position of the black stripe in Fig. 4.8a, so that it lies in the middle of the image rather than at the right side, has no effect on the Fourier spectrum so long as the width of the black and white areas remain identical. This is because there is no change to the pattern that is formed when the image is tiled infinitely. However, the *position* of the stripe is encoded in the complex data (images **c** and **d**). Note the difference between the real and imaginary data in this figure when compared with Fig. 4.8c and d

The difference between the coefficients of the real and imaginary data that we are interested in is called the *phase* angle ($\phi$):

$$\phi = tan^{-1}\left(\frac{I}{R}\right) \tag{4.10}$$

When a complex number is plotted as a vector in a complex plane diagram the phase angle is the angle between the positive real axis and the vector. The magnitude, what we show in the Fourier spectrum, is the length of the vector. See Appendix C for a graphical illustration.

The forward and inverse Fourier transforms can operate on complex data expressed in either real/imaginary format or magnitude/phase format. The two formats contain equivalent information and can be interconverted (Fig. 4.24).

When displaying representations of complex frequency domain data most of the time we can safely ignore the phase information and just look at Fourier spectra. This is not to say that phase data is unimportant and can be discarded. It cannot, but we *mostly* let the computer deal with it and don't bother to create displays of it. Figure 4.25 illustrates the effect of applying the inverse Fourier transform to only the magnitude data, or only the phase data when attempting to reconstruct Fig. 4.8a from its complex frequency domain equivalent.

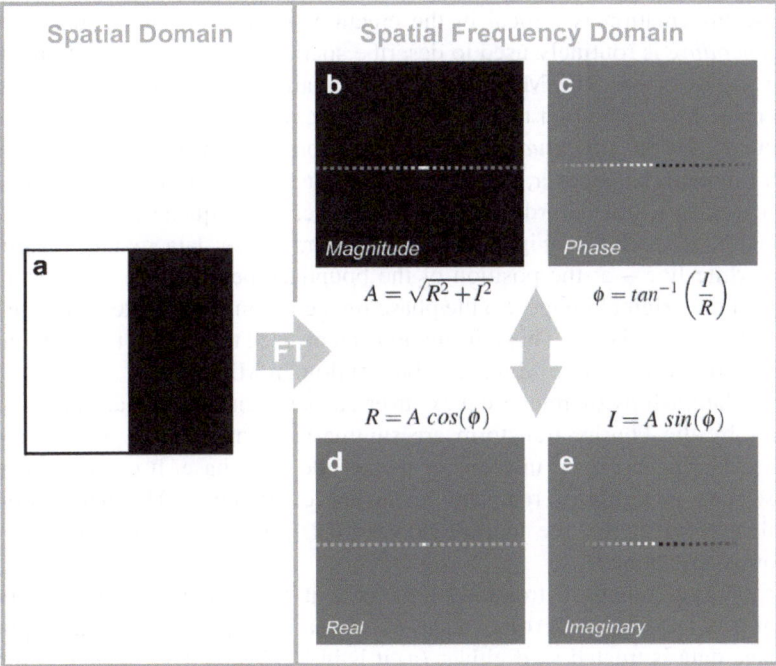

**Fig. 4.24** The forward and inverse Fourier transforms can operate on complex data expressed in either *Real + Imaginary* format, or *Magnitude + Phase* format. The two formats contain equivalent information and can be interconverted

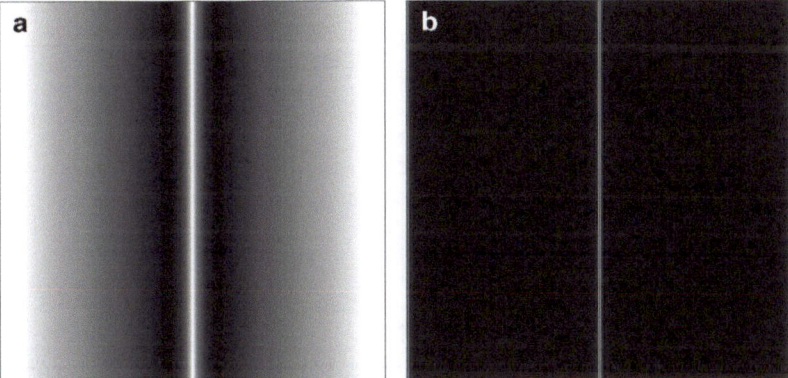

**Fig. 4.25** When an image is converted to its frequency domain equivalent by Fourier transformation both real and imaginary components are created. To recreate the spatial domain image by inverse Fourier transformation both the real and imaginary data are required. Image **a** here illustrates the result of inverse transformation of just the magnitude data from the Fourier transform of Fig. 4.8a. Image **b** illustrates the result of inverse transformation of just the phase data

Phase information is critical in the creation of magnetic resonance images – *phase encoding* is routinely used to describe spatial position in at least one dimension of an MR image. Raw MRI data always contains matrices of real and imaginary coefficients. Inverse Fourier transformation of the raw data creates *another* complex data set and it is the *magnitude* of this data that forms the spatial domain anatomical image. Remember, however, that although we display magnitude data this is *not* a Fourier spectrum – the coordinates now are space, not frequency.

Notice that the image Fig. 4.25b created from phase data mainly comprises a central white line – at the position of the boundary between the black and white regions in the original image. So the phase image is something like an image of the edges in the original data. This information can be used to identify tissue boundaries and measure the velocity of fluid (e.g. blood) flow in MRI.

The relationships of image data converted between the spatial and frequency domains by the Fourier transform are summarized in Fig. 4.26. In this diagram complex data is described in terms of magnitude and phase. It could alternatively be described in terms of real and imaginary coefficients. The descriptions are equivalent provided that the appropriate versions of the forward and inverse Fourier transforms are applied.

The Fourier transform treats *all* data as if it were complex. When a Fourier transform is used to convert an image to its spatial frequency domain equivalent the image data is treated as if either: (a) it is in Real/Imaginary format and all the

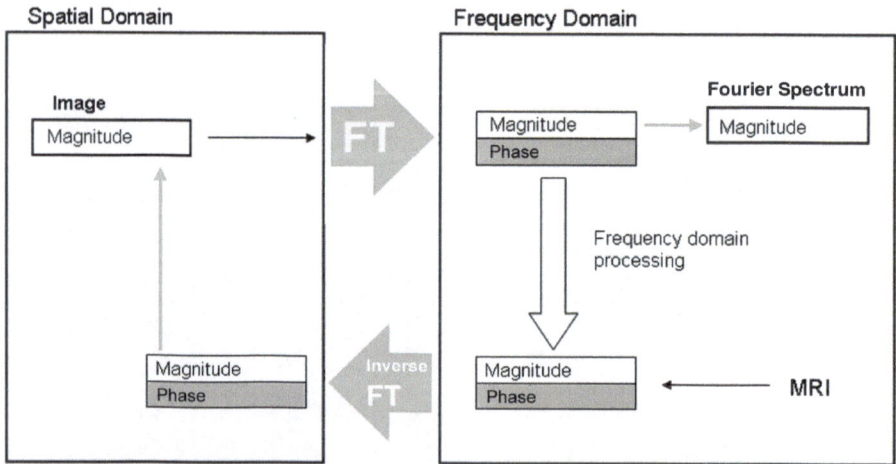

**Fig. 4.26** The relationships of image data converted between the spatial and frequency domains by the Fourier transform. A spatial domain image is comprised of magnitude (intensity) data only. Fourier transformation of this magnitude data produces a complex frequency domain data set that can be expressed in terms of magnitude and phase, as shown here, or real and imaginary coefficients (not shown). The Fourier spectrum is a display of the magnitude part of the frequency domain data. Inverse Fourier transformation of complex frequency domain data produces a complex spatial domain data set. We make an image from the magnitude of this complex data. In MRI raw data is acquired in the spatial frequency domain. An MR image is a magnitude display of the inverse Fourier transform of the raw data

imaginary coefficients are zero; or (b) it is in Magnitude/Phase format and all the phase angles are zero. In the former case the pixel intensities represent the real co-efficients, and in the latter they represent the magnitude. These two representations are in fact identical.

## 4.5 Two Practical Applications of Fourier Transforms

So far our discussions of the concept of spatial frequency, and transformations between the spatial and spatial frequency domain, have been illustrated with deliberately simple artificial images rather than medical images. We looked at simple images so that the discussion would not be sidetracked or confused by too many effects being present simultaneously. Now that we have a basic understanding of how the Fourier transform crocodile behaves, and hopefully it is a little less scary, we can look at what happens when we toss a medical image into the enclosure.

### 4.5.1 How Does the Focal Spot of an X-Ray Tube Affect Image Resolution?

One of the determinants of X-ray and CT image quality is the physical size of the source of the X-rays – the focal spot on the anode of the X-ray tube. A large focal spot will produce more blurry images than a small spot. A lot of engineering energy has gone into development of X-ray tubes that will produce very intense X-ray beams from the smallest possible focal spot. Nevertheless, since about 98% of the electron beam energy gets turned into anode heat rather than X-ray photons, there is always a compromise between small focal spot size and beam intensity. Many X-ray machines permit the adjustment of focal spot size according to the resolution and speed requirements of the imaging study. What effect does the size and shape of the focal spot have on the spatial resolution of an X-ray image? We can quantify the answer to this question by careful measurement and use of a Fourier transform.

We know the size and shape of the focal spot affect the sharpness of images but how do we measure this? We start with an image of the focal spot – easily produced with a pinhole camera. In this case the pinhole is a very small hole in a sheet of lead and we form the image on a high resolution sensor which could be a digital flat panel or a low speed film (without a fluorescent intensifying screen that would introduce extraneous blur). Suitably magnified we would expect to see the image looking something like Fig. 4.27a. The spot image is not a neat circular blob but an irregular blurred rectangle with two bright edges – images of the X-ray tube filament (a long coil of wire) that result from imperfect focusing of the electron beam. Since the spot image is a rectangle it should be intuitively obvious that the amount of image blur due to the finite size of the spot is going to be greater along one axis of the imaging plane that it is along the perpendicular axis.

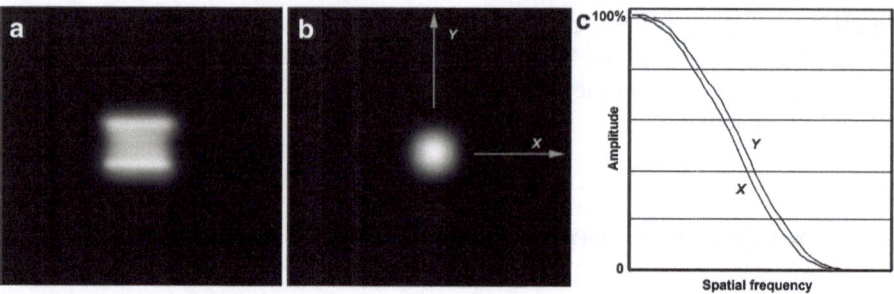

**Fig. 4.27** A pinhole image of the focal spot of an X-ray tube (**a**) and its Fourier spectrum (**b**). The profile of the Fourier spectrum (**c**) shows the way contrast is lost with increasing spatial frequency in the $x$ and $y$ directions. Low and medium spatial frequencies, represented by the bright center of the Fourier spectrum, have high amplitude and consequently good contrast. High spatial frequencies, represented by the dark outer areas of the Fourier spectrum, have low amplitude and consequently very poor contrast. Notice that the slightly asymmetrical shape of the Fourier spectrum is consistent with the 2D profile of the focal spot image. The bright horizontal bars in image **a** are, effectively, two images of the X-ray tube filament that result from imperfect focusing of the electron beam onto the anode of the X-ray tube

The blurring caused by the finite sized focal spot will blur every point in all imaged objects. Put another way, the image of every point in the imaged object will be a more or less bright spot having the same size and shape as the focal spot image. Small well-defined objects, which as we now know are described by a large range of spatial frequencies, will appear in the image with low contrast. The way to quantify the loss of contrast *for each spatial frequency* is to Fourier transform the image of the focal spot and examine the Fourier spectrum – Fig. 4.27b. The Fourier spectrum has a bright central region corresponding to high amplitudes for the low and medium spatial frequencies. So low and medium spatial frequencies will, in the absence of other blurring factors such as an intensifying screen, have high amplitudes in an image produced with this focal spot. All the high frequencies in the Fourier spectrum have very low amplitude. So high spatial frequencies will have very low contrast in an image produced with this focal spot – all small well-defined objects and all sharp edges will be blurred.

The profile of a line drawn from the center of the Fourier spectrum (Fig. 4.27c) tells us the degree of attenuation of spatial frequencies in the direction of the line. The profile would be symmetrical about the zero frequency point because the Fourier spectra of images (2D arrays of real numbers) are always symmetrical. The shape of the profile of the line is called the *Modulation Transfer Function*. The MTF is used extensively in quantitative description of imaging system performance and we will discuss it more fully in Chapter 5.

## 4.5.2 Making Diagnostic Images from Raw MRI Data

Finally, lets look at MRI data, which is directly acquired in the spatial frequency domain. Figure 4.28 shows the Fourier spectrum of raw complex data from a

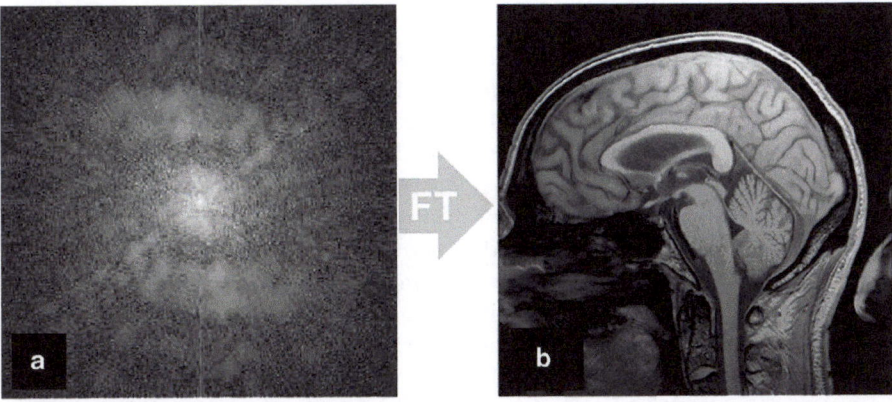

**Fig. 4.28** Fourier spectrum of raw frequency domain MRI data (**a**), and the spatial domain diagnostic image (**b**) formed by inverse Fourier transformation

scan of a human brain. The magnitude of the inverse Fourier transform gives us the diagnostic spatial domain image.

Although the Fourier spectrum looks rather complicated we can interpret some of its features in terms of the resultant spatial domain image. We see plenty of signal at all spatial frequencies, consistent with the presence of both edge and tonal detail in the spatial domain image. There are a number of weak radial lines in the Fourier spectrum encoding several distinct edges in the anatomical image – the back of the skull particularly. The narrow vertical line is consistent with the intensity difference between the top and bottom edges of the image.

There is another particularly interesting feature of the spatial domain image. The subject's nose, which is just outside the field of view on the left side of the image, has appeared behind his head! This artifact, called phase wrap, occurs because the signal from the nose, although outside the field of view, is nevertheless detected and recorded in the complex frequency domain raw data. It appears on the opposite side of the image because of the effective tiling of the frequency domain data.

## 4.6 Summary

- A digital image is a *spatial domain* data set. The coordinates of the image data represent distance or position in space.
- The term spatial frequency describes the rate of change of intensity in an image in terms of a sinusoidal intensity profile. Any intensity profile can be described as the sum of a series of sinusoids of appropriate spatial frequency and amplitude. Any image can be described as the sum of a series of $x$ and $y$ direction sinusoids.

- A 2D Fourier transform converts an image into its spatial frequency equivalent in the *spatial frequency domain*. The Fourier transform converts the *real* intensity values of a spatial domain image into a symmetrical *complex* frequency domain matrix. The Fourier transform of an image is an *exact* representation of the image. No information is lost in the process of Fourier transformation.
- Complex data can be expressed in two equivalent interconvertible forms: *Real* and *Imaginary* coefficients; or *Magnitude* and *Phase* coefficients.
- The *Fourier spectrum* of a digital image is a display of the *magnitude* of the spatial frequency domain version of the image data. The coordinates of the spectrum data represent differences in spatial frequency.
- By convention, 2D Fourier spectra are displayed with the lower frequency terms in the center of the spectrum. In the middle of the spectrum is the 'zero frequency' or 'DC' term. The magnitude or amplitude of the DC term represents the average pixel intensity in the spatial domain image. Because the amplitude of the DC term is normally much greater than the amplitude of the non-zero frequencies, it is normal to display the log of the amplitudes in the Fourier spectrum.
- Lines and sharp edges in images are characterized by non-zero amplitudes of many spatial frequencies. Straight lines and edges in images give rise to linear features in the Fourier spectrum, and these features lie perpendicular to their originating lines in the spatial domain image.
- The Fourier transform effectively treats 2D data (for example an image) as if it were tiled infinitely in space. Thus discontinuities in intensity at the joins between edges of the original data will give rise to linear features in the Fourier spectrum that are oriented perpendicular to the originating edges.
- An image that contains no sharp edge detail (and no edge discontinuities when tiled) can be described with low spatial frequencies only. The Fourier spectrum of such an image will have non-zero values only near its center.
- Inverse Fourier transformation of complex spatial frequency domain data produces complex spatial domain data. We normally view only the magnitude of this complex data. Raw MRI data is an example of a complex spatial frequency domain data set.

# Chapter 5
# Image Quality

## 5.1 Introduction

Most of us can look at an image and immediately make some sort of judgement about its quality. We might notice that the image is blurry, or the colors are unnatural, or we can't read some text in the image because it is almost the same color as the background. These *subjective* assessments of image quality are made in relation to either the purposes of the viewer or, possibly as a secondary thought, the purposes intended by the image's creator. In visual art, at least, there may be a huge intellectual gap between those two purposes. In the case of medical images, however, the purposes of the image creator and viewer should be identical and well-defined. The purpose of a medical image is to provide information about *specific* aspects of body structure or function. The information in the image, in other words the *quality* of the image, must be sufficient to permit an accurate assessment of the structures or functions in question. If a doctor requests an X-ray for a suspected fracture then we want an image that will reveal not just the presence of a fracture but also its extent and severity. Such information may affect the way the patient's injury is managed. The radiologist reviewing the images will have very firm ideas about what constitutes a good quality image for the purpose of detecting and describing a bone fracture.

Since the purposes of medical images are objectively well-defined it is desirable to have quantitative methods of describing image quality. These methods are essential to the objective comparison of images and imaging systems, and the optimization of imaging technique and system design. Technique optimization is especially important when, as in X-ray imaging, there is an essential conflict between image quality and potential harm to the patient. Imaging cost is also an important consideration. Medical imaging systems are expensive to buy and operate so it is important to produce the best quality images in the shortest possible time.

In general we can say that the most important factors contributing to medical image quality are *Contrast*, *Spatial Resolution*, and *Noise*. *Ideally* we would like high contrast, high spatial resolution, and low noise, however, these are not independent factors – they affect each other in complex and confusing ways. If either contrast or spatial resolution is too low, or if noise is too high then, as illustrated in Fig. 5.1, the

R. Bourne, *Fundamentals of Digital Imaging in Medicine*,
DOI 10.1007/978-1-84882-087-6_5, © Springer-Verlag London Limited 2010

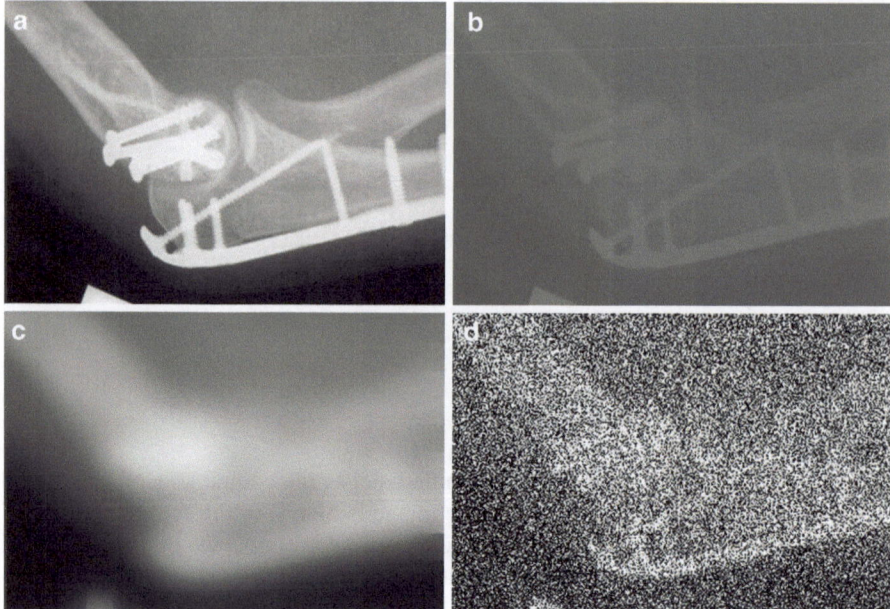

**Fig. 5.1** A useful image must have adequate contrast and resolution, and a low noise level, as illustrated in the image (**a**). Image **b** has high spatial resolution and low noise, but is rendered useless by having almost zero contrast. Image **c** has low noise and high contrast, but extremely poor spatial resolution. In image **d** we see high spatial resolution but the very high noise level has destroyed the contrast information

image is of no value – no clinical information can be extracted. The actual levels of contrast and resolution that we need, and noise that we can tolerate, depends on the purpose of the image. They also depend on innate properties of the object imaged (mostly tissue), the properties of the imaging system (hardware), and the way the system is used (technique).

In this chapter we will first discuss contrast and noise separately and then, because noise strongly affects contrast, look at them together. We will discuss spatial resolution last because it is affected by both contrast and noise.

## 5.2  Contrast

All imaging techniques depend on *differential* emission of energy from the imaged object according to some physical property of the object. Without differential emission the measured signal would contain no information about the imaged object and the only variations would be informationless noise. *Contrast* is a measure of the magnitude of the measured signal differences between physically different regions of the imaged object. When these measured signals are converted into an image 'contrast' describes the magnitude of intensity differences between different regions in the image. We can thus think of image contrast as being the product of *signal contrast* and *detector contrast*:

$$C_I = C_S \times C_D \tag{5.1}$$

The signal contrast $(C_S)$ depends on the energy source and the physical properties of the imaged object – it describes the range of energies emitted by the object. The detector contrast $(C_D)$ depends on the way the signal emitted by the imaged object is modified (e.g. with a scatter suppression grid), detected, and recorded. We need *both* signal and detector contrast to form image contrast.

In digital imaging it is very easy to enhance or reduce image intensity differences in order to make them more or less obvious to a human observer. This is the process of *contrast adjustment*. However, there is no way to enhance contrast if there is no difference in the measured signal or raw data. Also, a simple contrast adjustment that exaggerates (amplifies) recorded signal differences will also exaggerate any noise. The result may be no improvement in the ability to extract information from the image.

### 5.2.1 Simple Measures of Contrast

Figure 5.2 shows two version of a wrist MR image. Image a plainly has higher contrast than image b – the bones are much brighter *relative to the surrounding tissue background*, even though in image b the average brightness of the bones is greater than in image a. How could we measure these contrast differences *quantitatively*?

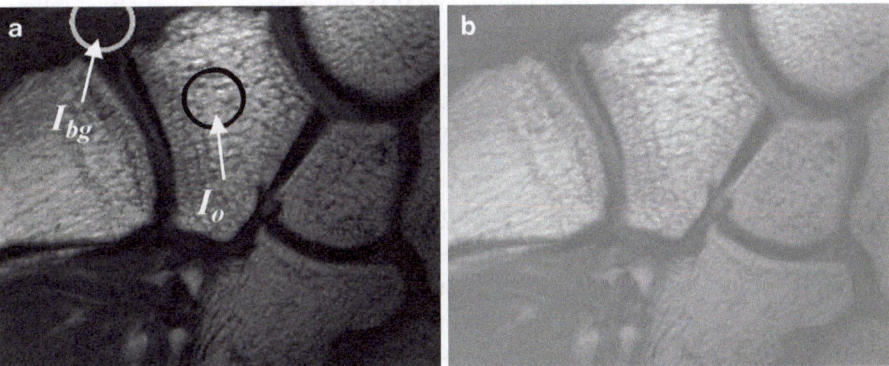

**Fig. 5.2** Two versions of a wrist MR image. Image **a** plainly has higher contrast than image **b** – the bones are much brighter relative to the surrounding tissue background, even though in image **b** the average brightness of the bones is greater than in image **a**. Measures of contrast are based on the relative or absolute difference in average intensity of an object and its background

A simple way to do this would be to measure the average pixel intensity in a 'typical' bone region (inside the black circle, say) and in a typical soft tissue 'background' region (inside the white circle). The difference between these average intensities is 124 for image a and 63 for image b. This measurement suggests that the contrast in image a is about twice the contrast in image b. This figure seems a little out of kilter with our perception which would suggest *much* greater contrast difference between the two images. Another problem with this measurement is that we don't have a meaningful measure of the contrast in a single image. The intensity difference of 124 measured from image a is based on an 8 bit scale ranging from 0-255. If we had a 10 or 12 bit image we would get different measures for the intensity difference.

A more general measure of contrast describes the intensity difference relative to the background:

$$C = \frac{I_o - I_{bg}}{I_{bg}} \tag{5.2}$$

where $I_o$ and $I_{bg}$ are the average pixel intensities in the object and its background.

Using this measure we would get contrast values for bone versus background of 3.35 for image a and 0.44 for image b, suggesting that the contrast difference in image a is 7.6 times greater than in image b. This measure perhaps exaggerates the difference relative to our perception. Also, sometimes it is an arbitrary choice as to which region is the 'object' and which the background, yet the choice may make a significant difference to the contrast calculated by Eq. 5.2. We can avoid this problem by using the expression:

$$C = \frac{I_o - I_{bg}}{I_o + I_{bg}} \tag{5.3}$$

giving contrast values 0.63 for image a and 0.18 for image b, or a contrast ratio of 3.5 which is in good agreement with perception. Using this expression the choice of background and object affects only the sign of the calculated contrast. We will see soon that this expression is very similar to the definition of signal 'modulation'.

Another common perception-related way to measure contrast is the log of the intensity *ratio*:

$$C = log_{10}\frac{I_o}{I_{bg}} \tag{5.4}$$

For a film image this definition of contrast is identical to the *optical density* difference $(OD_o - OD_{bg})$. Human visual perception of intensity is non-linear (see Fig. 6.20) so this logarithmic scaling makes sense. For Fig. 5.2 it gives bone to soft tissue contrasts of 0.62 for image a and 0.16 for image b, or a 3.9 times difference between images a and b – a reasonably close match with our perception.

## 5.2.2 Contrast and Spatial Frequency

Human perception of contrast is dependent on spatial frequency. For a sinusoidal contrast pattern (Fig. 5.3) maximum contrast sensitivity occurs at a spatial frequency around three cycles per degree. Here a degree refers to an angle in the visual field. If we view an image at a distance of 400 mm, then one degree is 7 mm, so three cycles per degree = one cycle per 2.33 mm. Contrast sensitivity is about four times lower at frequencies of 0.5 and 8 cycles per degree, and 100 times lower at 40 cycles per degree.

## 5.2.3 Optimizing Contrast

The choice of imaging modality and imaging technique are fundamentally decisions about contrast because contrast is the image property that contains information. Spatial resolution describes the bottom limit of the scale of information potentially available, but as Fig. 5.1 clearly demonstrates, high spatial resolution is meaningless in the absence of contrast. If contrast in the measured signal cannot be obtained by virtue of the inherent physical properties of the imaged tissue and selection of an appropriate energy source, then it is often added artificially. A radiographer 'gives contrast' by injecting, or getting the patient to drink, a modality-specific contrast agent – an X-ray absorbing metal such as barium, a paramagnetic metal for MRI, or 'microbubbles' for ultrasound imaging.

**Fig. 5.3** Illustration of the spatial frequency dependence of contrast sensitivity. Contrast sensitivity is greatest at a spatial frequency of about 3 cycles per degree of visual field ($\approx$1 cycle/3 mm at a viewing distance of 500 mm). Viewing this image at different distances will demonstrate the effect – the highest point of the bar pattern will move to the right as viewing distance decreases

## 5.3 Image Noise

No imaging method works without contrast and no imaging method is free of noise. If contrast is low and noise is high then the random intensity variations due to noise will make it difficult to visually detect the intensity changes due to contrast. Figure 5.4 illustrates how an increasing level of noise relative to contrast diminishes our ability to distinguish objects in an image. But before discussing the interrelation between noise and contrast we will first consider the origin and types of noise found in medical images.

### 5.3.1 What Is Noise?

The everyday answer to this question is *'Annoying sounds that make it difficult to hear what you want to hear'*. In terms of imaging a similarly broad definition would be *'Any intensity or color fluctuations that make it difficult to see what you want to see'*. The problem with these definitions is that we often don't *know* what it is we are supposed to be hearing or seeing – some or even all the information is hidden by the noise. What we want ideally is to be confident about the information we

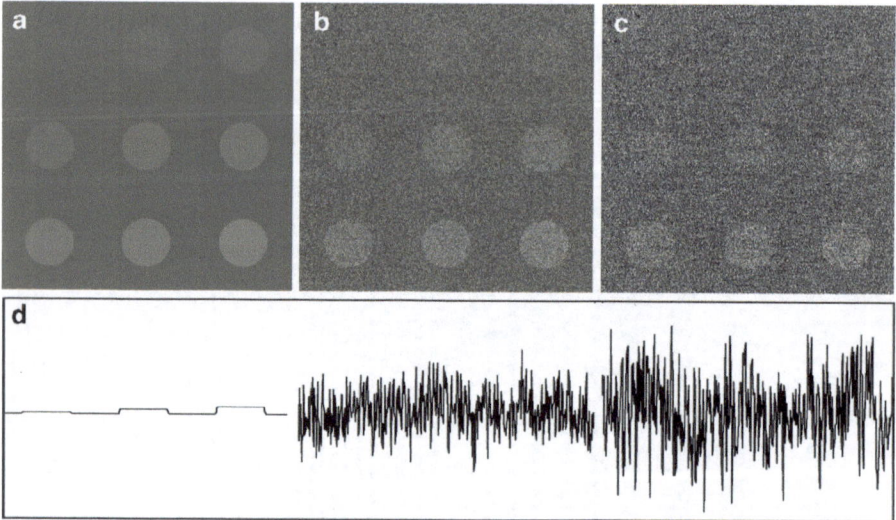

**Fig. 5.4** The presence of noise reduces our ability to extract information from an image. The noise-free test pattern (**a**) contains nine circles of equal diameter with contrast ranging from 0.003 to 0.069 according to Eq. 5.4. In the presence of noise (images **b** and **c**) the visibility of the lower contrast objects is reduced or completely lost. Plot **d** shows the intensity profile through the top row of circles. Notice that objects are easier to distinguish in the image than in the intensity profile. This is because our perception computes a spatial average of intensities in the (2D) image but not in the 1D profile

receive. If you were on a very noisy phone connection you might ask your caller to repeat a sentence three times before you are sure you know all the words in the sentence. You may never hear one sentence fully due to the noise but so long as the caller repeats the same sentence each time you eventually know what all the words are. When we make an image we can effectively do the same thing to reduce the uncertainties due to noise – we either measure the signal for a longer time, or we repeat the measurement several times, which is effectively the same. If the noise is random its contribution to the total image signal will diminish over time or repeated measurements.

What happens if the noise is not random? This type of noise does not decrease when the measured signal is averaged over time. It can appear due to leakage of a spurious signal into the system (from the mains power supply for example), or some systematic 'error' in the method of formation of the image. The latter type of signal is generally referred to as an *artifact*.

Focusing on imaging we can, at least conceptually, separate the energy we measure, the image ($I$), into 'true' signal variations that are characteristic of the imaged object ($S$), and 'spurious' or 'random' variations that are *not* characteristic of the imaged object – in other words noise ($N$) (and possibly artifacts but we will ignore these for now):

$$I = S + N \tag{5.5}$$

While we would generally think of the 'signal', $S$, as always being positive, while the noise, $N$, could be expected to make either positive, zero, or negative contributions to a measurement. The 'erroneous' measurement value it causes may be either higher or lower than the 'true' signal measurement expected. In many cases we can say that the average (mean) measurement error due to noise is zero. This is 'zero-mean' noise.

In general we have no way of separating 'true signal' from 'random noise'. Nevertheless, the concept of separable signal and noise is very useful in signal and image processing. This does, however, cause some potentially confusing terminology. When we talk about the intensity of the energy we measure or 'detect' this is usually referred to as 'the signal' – *'I'm not getting a signal!'* or *'The signal is strong enough, but it's very noisy'*. A few seconds later somebody will ask, *'What's the signal-to-noise ratio?'* Suddenly the meaning of 'signal' has changed. In discussions that include noise, 'signal' usually means the notional 'true' signal, free of noise, coming from the object of interest. If the discussion does not explicitly mention noise then 'signal' probably means the intensity of detected energy, including noise. Most of the time the context is sufficient to work out which usage is applicable.

## 5.3.2 Quantum Mottle

Going back to our general definition of image noise as 'Any intensity or color fluctuations that make it difficult to see what you want to see' we should include in this

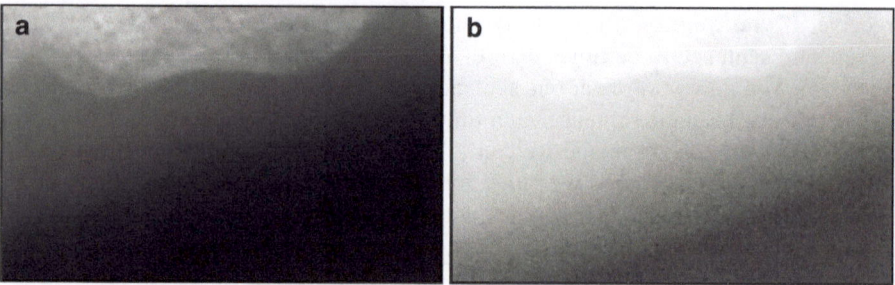

**Fig. 5.5** The random 'grainy' pattern in the background of a radiograph (**a**) is called *Quantum mottle*. Here it is made more obvious in image (**b**) by contrast enhancement of the darker pixels. Quantum mottle is the natural result of statistical variations in the number of X-ray photons incident on a particular part of the detector in a finite measurement time interval

random fluctuations in the flow and spatial distribution of energy from the energy source itself. In the majority of medical imaging methods the energy source directly (or, in the case of PET, indirectly) emits electromagnetic radiation, or *photons*. The particulate nature of electromagnetic radiation has a profound effect on the nature of noise in these medical images. The probability of a photon being detected in any region of the image sensor during any specific interval of time is *constant*. We can see this effect if we look closely at the background of a radiograph (Fig. 5.5) where the variation in measured intensity is due entirely to the (locally) random distribution of photons emitted by the X-ray source. This random 'texture' is called *Quantum mottle*.

Quantum mottle is the result of a *Poisson* process, meaning that if we measure the standard deviation ($\sigma$) in the signal for all identical sensor regions (pixels in the case of a digital sensor) it will be equal to the square root of the average signal ($\mu$). It is helpful to describe this variation in relative terms:

$$Relative\ variation = \frac{\sqrt{\mu}}{\mu} = \frac{1}{\sqrt{\mu}} \tag{5.6}$$

The significance of this expression is that the more accumulated signal that we get, the smaller is the relative variation. In other words, quantum mottle *decreases*, relative to total intensity, with increasing measurement duration or signal energy flux. If we measure for four times as long, or take the average of four measurements, then the amount of quantum mottle noise will decrease by one half. Poisson statistics is applicable to a large number of physical processes occurring in medical imaging.

### 5.3.3  Other Noises

Quantum mottle is an example of 'measurement noise' that is an inevitable consequence of the use of electromagnetic radiation as the imaging energy source. Even when we have reliable statistics in the detected energy there are other ways noise can enter the recorded signal and thus appear in an image. The most obvious of these noise sources is thermal noise in the sample and the electronic hardware associated with signal detection, amplification, and recording. Also, spurious electromagnetic radiation from external sources can leak into the analog electronics. The digital components of an imaging system are much less susceptible to external noise but can nevertheless give rise to random variations in reconstructed images due to lack of calculation precision.

### 5.3.3.1 Gaussian Noise

Random noise that enters the imaging system from external sources typically has a Gaussian, or normal, distribution. The magnitude of the 'error', or uncertainty, in the recorded signal due to this type of noise is not dependent on the total signal, as is the case for quantum mottle. In an image Gaussian noise affects dark and light areas to the same degree (Fig. 5.6a).

The intensity uncertainty due to Gaussian noise can be described by a *Probability Distribution Function* (PDF) shown as the familiar 'bell curve' in Fig. 5.6b. In imaging terms this PDF tells us there is a high probability that, for any particular pixel, the error due to noise will be small. Large errors have very low probability. The average of all errors is zero because the noise can either increase or decrease the recorded signal. The standard deviation ($\sigma$) of the PDF gives an indication of the average error *magnitudes*. For a Gaussian distribution $\frac{2}{3}$ of all pixels will have an error less than ($\sigma$). It is important to remember, however, that although we may be able to summarize the error statistics there is no way to know how big the error is for any particular individual measurement (pixel). Thus there is no way to completely remove the noise.

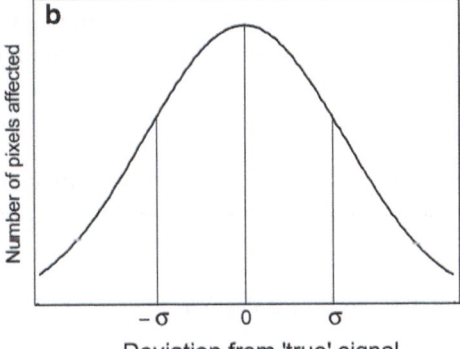

**Fig. 5.6** Gaussian noise affects light and dark areas of an image equally. Image **a** shows a test pattern without and with Gaussian noise. The *Probability Distribution Function* (PDF) (**b**) describes the distribution of the noise errors. The standard deviation ($\sigma$) gives a summary of the range of error magnitudes. For an image with a Gaussian noise distribution, $\frac{2}{3}$ of all pixels will have an error less than ($\sigma$). A decrease in Gaussian noise is equivalent to a decrease in $\sigma$

Gaussian noise is sometimes called 'additive noise' as its contribution to an image, $I$, can be modeled as a simple addition of noise having a Gaussian PDF, $N_G$, to the noiseless signal, $S$:

$$I = S + N_G \tag{5.7}$$

### 5.3.3.2 Speckle

Speckle is the dominant form of noise in medical ultrasound imaging (Fig. 3.14) because many microscopic tissue components, smaller than the spatial resolution of the technique, act to reflect sound waves incoherently. When waves of a single wavelength are reflected from a rough surface each point on the surface effectively acts a source of spherical waves. The roughness of the surface means that these numerous spherical waves will have many different phases which will cause the waves to interact *constructively* at some points in space and *destructively* at other points. If the surface is rough enough to cause phase differences greater than $2\pi$ (360°) then the resultant intensity pattern will be random. The maximum possible intensity would be the sum of the amplitudes of a large number of large amplitude spherical waves, whereas the minimum intensity, when all waves interact destructively, would be zero. High intensity constructive interactions are much more probable in regions where numerous high amplitude spherical waves are present – in other words, in regions of high average signal intensity. In regions of low average signal intensity constructive interactions will produce only medium or low amplitudes. The overall result is that the random (noise) signal will have an average amplitude that increases with overall signal intensity. In an image this appears as many bright specks in the lighter regions. This is *Speckle* noise.

Speckle is not an *entirely* bad thing in ultrasound. The speckle pattern is not always completely random – in which case it contains information about the tissue. Some pathologies give rise to a characteristic speckle signal. In contrast to Gaussian 'additive' noise, speckle noise is called 'multiplicative noise'. It's contribution to an image, $I$, can be modeled as multiplication of the noiseless signal, $S$, by random numbers having a zero-mean Gaussian PDF, $N_G$:

$$I = S + (S \times N_G) \tag{5.8}$$

Note that in this model we are still adding the noise to the 'true' signal, but this time the magnitude of the noise is proportional to the signal intensity.

### 5.3.3.3 Salt and Pepper

Most image noise causes relatively small random deviations from the expected 'true' image intensity. Extreme intensity deviations may also occur resulting in image pixels being incorrectly assigned the maximum or minimum possible values. This 'Salt

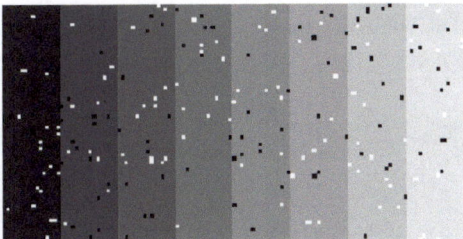

**Fig. 5.7** 'Salt and Pepper', or *Impulse*, noise appears as randomly distributed *white* and *black* pixels in a digital image

and Pepper', or *Impulse*, noise appears as randomly distributed white and black pixels in a digital image (Fig. 5.7). It can arise from defects in single elements of a semiconductor sensor, sometimes called 'stuck pixels', defects in a semiconductor memory or storage device, noise affecting data transmission, or even image reconstruction and processing errors.

#### 5.3.3.4 Artifacts

Our broad definition of image noise as *'Any intensity or color fluctuations that make it difficult to see what you want to see'* would include systematic intensity errors that occur due to specific properties of the imaging method, or the interaction of these properties with those of the imaged object. Systematic errors of this kind are referred to as *artifacts*. We previously mentioned the 'chemical shift artifact' in MRI that causes the image of fat to be displaced relative to adjacent tissue. Because the origin of artifacts is highly specific to the imaging modality, rather than a general characteristic of the flow and differential transmission of imaging energy, the solutions are also very modality specific and we will not investigate them further in this text.

### 5.3.4 Signal to Noise Ratio

As we saw in Fig. 5.4 noise can have a significant effect on our ability to see objects in images. A common way to quantify the level of noise in an image (or any measured signal) is to estimate the *Signal-to-Noise Ratio* (SNR):

$$SNR = \frac{S}{N} \tag{5.9}$$

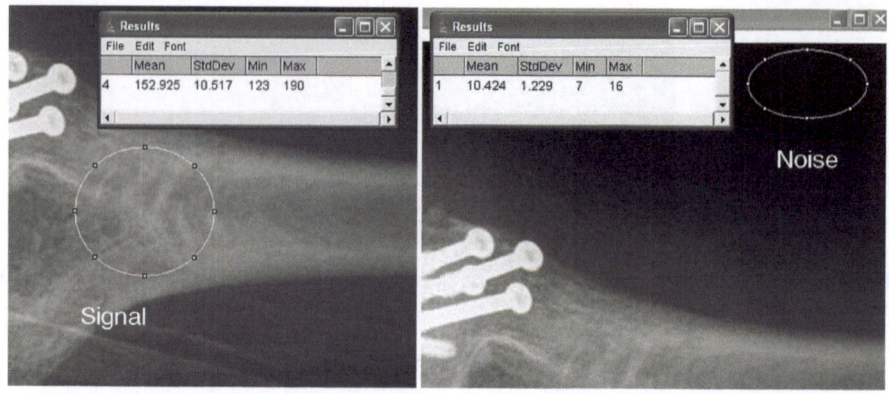

**Fig. 5.8** A simple estimate of *Signal to Noise Ratio* (SNR) in an X-ray image. The *Signal* is measured as the average intensity in a selected region of the anatomy, in this case the base of the humerus, and the *Noise* in a region of the image where there is no signal from anatomy. We can then calculate (using Eq. 5.10) that, for the base of the humerus, the SNR is $0.66 \times \frac{152.9}{1.23} = 77.6$. Obviously, this method of measuring SNR is highly dependent on where in the image the signal is measured and it takes no account of the importance of contrast (the 'Results' windows are the output of the ImageJ Menu: Analyze > Measure command applied after drawing the ROI)

In this expression we would not want to use the mean value of the noise as, at least for zero-mean Gaussian noise, this would give an inappropriate infinite SNR value. Instead it is conventional to use the standard deviation of a tissue-free region in the background($\sigma_{BG}$) as an estimate of the noise. The 'signal' can be taken as the mean intensity ($\mu_S$) in the anatomy of interest (Fig. 5.8):

$$SNR = 0.66 \times \frac{\mu_S}{\sigma_{BG}} \qquad (5.10)$$

A 'correction factor' of 0.66 is applied because in the background the 'negative' part of the zero-mean noise appears black and the measured standard deviation is less than the true value expected for a Gaussian PDF.

The 'difference method' is useful for measuring SNR of an imaging system rather than the SNR of a single image. In this method two images ($I_1$ and $I_2$) of a test object are acquired one after another without changing any imaging parameters. The 'signal' is then taken as the mean intensity ($\mu_S$) in a ROI placed inside the image of the test object. The 'noise' is measured from the standard deviation ($\sigma_D$) in the same ROI in the *difference* image ($D = I_1 - I_2$). Notice that the difference image can have negative pixel intensities. The SNR is then calculated using the expression:

$$SNR = \sqrt{2} \times \frac{\mu_S}{\sigma_D} \qquad (5.11)$$

In this case the correction factor compensates for the exaggeration of noise in the difference image.

## 5.4   Contrast + Noise

The problem with image SNR estimates is that they reveal nothing about the effect of noise on our ability to see objects in an image because visibility depends on contrast – the difference between signals. A highly overexposed radiograph might have a very high SNR and yet contain no useful information about the imaged object. A more useful estimate of the effect of noise on image information is the *Contrast-to-Noise Ratio* (CNR):

$$CNR = \frac{(\mu_A - \mu_B)}{\sigma_{BG}} \qquad (5.12)$$

In this expression noise is measured as the standard deviation of the background and the contrast measure is simply the intensity difference between an object and its background.

Figure 5.9 shows a series of nine circles of increasing contrast relative to a mid-gray background. The contrast values range from 0.004 to 0.086 (expressed here as a fraction of the possible intensity range to eliminate the significance of bit depth), while the standard deviation of the noise is 0.173. This gives us CNR values ranging from 0.02 to 0.50. We can perhaps just barely see the circle at the left end of the middle row (CNR = 0.35).

The ease of visual detection of an object in an image, sometimes called 'conspicuity', depends not just on the CNR but also on the size of the object. Figure 5.10a shows a series of various size circles of constant CNR relative to the background. The largest circle is easy to see but the smallest, at the top left is effectively invisible. Because our perception performs a local 2D averaging of intensities larger objects are generally easier to see than smaller ones. The visibility or conspicuity of an object is roughly proportional to its area. However, for the same area, more compact objects (such as circles) will be more visible than less compact objects (such as stars) as demonstrated in Fig. 5.10b.

**Fig. 5.9**  Illustration of the effect of noise and contrast on object visibility. The image contains nine circles of equal diameter with contrast values ranging from 0.004 (*top left*) to 0.086 (*bottom right*), according to Eq. 5.3. The standard deviation of the noise is 17% of the possible intensity range. This gives CNR values ranging from 0.02 to 0.50. We can perhaps just see the circle at the left end of the middle row (CNR = 0.35)

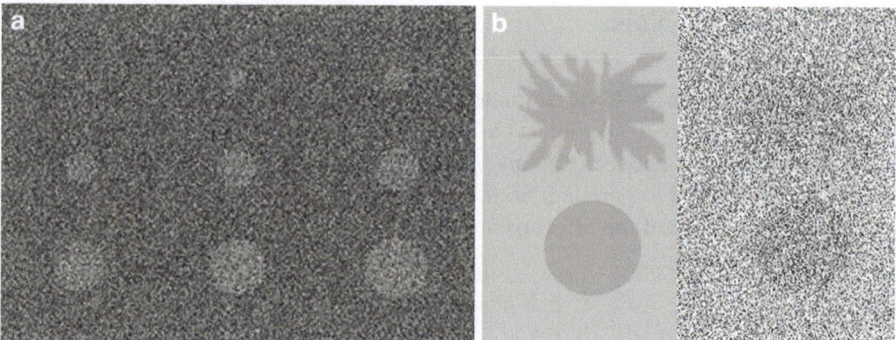

**Fig. 5.10** Illustration of the effect of object size and shape on visibility in the presence of noise. The test image **a** contains nine circles of equal contrast but a range of diameters. Because our perception performs a local 2D averaging of intensities, larger objects are generally easier to see than smaller ones. However, for objects having the same area (e.g. the circle and the 'squashed butterfly' in image **b**), a more compact object (such as a circle) will be more visible than a less compact object

It should now be clear that contrast, noise levels, and object size and shape all affect our ability to extract visual information from an image. There are several ways of measuring SNR and CNR so we need to be careful to specify the method used when communicating such measurements.

## 5.5   Spatial Resolution

In an ideal situation 'spatial resolution' describes the size of the smallest objects that can be separately discriminated by an imaging system. However, as we have just seen, the effects of contrast and noise are very significant, so a proper description of spatial resolution needs to incorporate these factors.

### 5.5.1   Line Pairs

In plain X-ray imaging it is common to test and measure the spatial resolution of a system in terms of *Line pairs* per millimeter (or centimeter). The line pairs test typically uses a test object which comprises a series of parallel lines of very high inherent contrast (Fig. 5.11a). For a plain X-ray imaging the test object is a thin metal plate in which a series of slots of progressively decreasing width and spacing are cut. One 'line pair' in an image is then one white line immediately adjacent to one black line.

When a line pairs test object (phantom) is imaged in an imperfect system the acquired image is blurred to some degree (Fig. 5.11b). A crude measure of the spatial resolution of the imaging system is the maximum spatial density of line pairs that

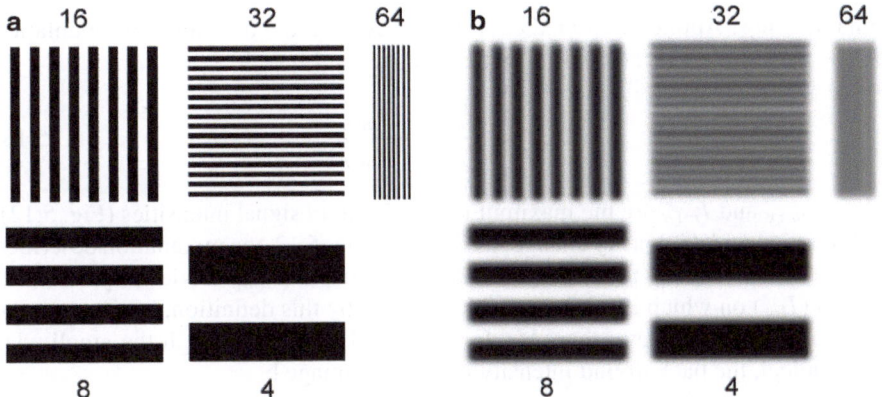

**Fig. 5.11** A line pairs test pattern (**a**) as typically used to measure the spatial resolution of a projection imaging device. The numbers specify the number of line pairs per unit distance. Note the similarity to the black and white stripe patterns of Figs 4.10 and 4.11. Each line pairs pattern is described by a primary sinusoidal spatial frequency identical to the line pairs frequency. Higher spatial frequencies are also present but at lower amplitude than the primary. Image **b** shows a possible image obtained by imaging the test object in an imperfect system. Note that the effect of imperfect spatial resolution is a loss of contrast at higher spatial frequency

can just be distinguished. This will correspond, very roughly, to the smallest width *linear* object that would be visible in the image. Spatial averaging by our perception makes lines more visible that points of the same dimension. Point objects of the same diameter as the minimum visible line width will be *invisible*. Notice in Fig. 5.11b that blurring causes a progressive loss of contrast as the line pairs density increases. Even at line pair densities above the limit of visual resolution the light part of the pattern is no longer white and the dark part is not black – decreasing spatial resolution is due to a *loss of contrast*.

It should be clear from Fig. 5.11b that visual inspection of a line pairs test image is not an ideal measure of spatial resolution as there is no clear point at which separate lines become indistinguishable, and there is significant loss of contrast for objects much larger than the visibility limit. The effect of the blurring is a general loss of edge and line detail, and a progressively more severe loss of contrast with increasing spatial frequency. The *inherent* very high contrast of the test object is not fully reproduced by the imaging system – especially at high spatial frequency.

## 5.5.2 The Modulation Transfer Function

Rather than attempt to define spatial resolution with a single 'minimum object size' parameter it is better to fully describe the way the inherent contrast of an imaged object is lost in the imaging system as spatial frequency increases. This is commonly done with the *Modulation Transfer Function* or MTF. MTF descriptions of imaging system performance are used widely – in microscopy, photography, and astronomy, to name a few. Applied to a medical imaging system, the definition of modulation ($M$) is very similar to the contrast definition used in Eq. 5.3:

$$M = \frac{I_{max} - I_{min}}{I_{max} + I_{min}} \qquad (5.13)$$

where $I_{max}$ and $I_{min}$ are the maximum and minimum signal intensities (Fig. 5.12).

For a sinusoidal intensity change, as shown in Fig. 5.12, the modulation described by Eq. 5.13 is identical to the amplitude of the sinusoid ($I_{amp}$) divided by the background ($I_{bg}$) on which it is imposed. Notice that, by this definition, for image b the modulation $M_B$ is greater than $M_A$ for image a because, although the amplitudes are identical, the background intensity is lower in image b.

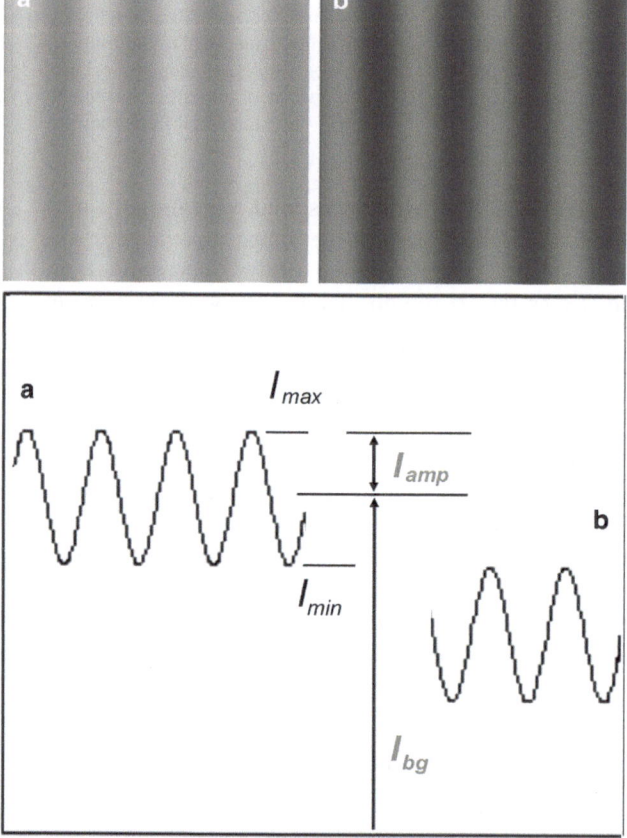

Fig. 5.12 The modulation ($M$) of image intensity is defined as $M = (I_{max} - I_{min})/(I_{max} + I_{min})$. For a sinusoidal intensity profile this definition is identical to $M = I_{amp}/I_{bg}$ where $I_{amp}$ is the amplitude of the sinusoid and $I_{bg}$ is the background intensity. Notice that for image **b** the modulation $M_B$ is greater than $M_A$ for image **a** because, although the amplitudes are identical, the background intensity is lower in image **b**

The MTF for an imaging system, or a component of an imaging system, is defined as:

$$MTF(f) = \frac{M_{out}(f)}{M_{in}(f)} \qquad (5.14)$$

where $M_{in}(f)$ is the 'input' signal modulation and $M_{out}(f)$ is the output signal modulation. The symbol $(f)$ is included to emphasize that we are interested in the way the input and output signal modulation changes with spatial frequency. When describing the MTF of a whole imaging system, $M_{in}$ would be the inherent contrast in the imaged object, and $M_{out}$ would be the image contrast.

For our notional imaging system that produced the blurred image shown in Fig. 5.11b we can estimate $MTF(f)$ by examination of the intensity profiles of the original and blurred line pairs patterns (Fig. 5.13). The intensity ($M_{in}(f)$) of the original pattern always varies between a maximum ($I_{max}$) of 255 and a minimum ($I_{min}$) of zero (this is an 8 bit image), so, according to Eq. 5.13, $M_{in}(f) = 1$ at *all* the line pair densities in the test pattern. $M_{out}(f)$, however, decreases with increasing spatial frequency. The five values of $M_{out}(f)$ at the spatial frequencies 4, 8, 16, 32, and 64 are approximately 1, 1, 0.8, 0.1, and 0 (we omit the space unit here as it depends on the final print size of the image). Notice that at line pair density 8 the intensity profile is quite rounded but $I_{max}$ and $I_{min}$ are still 255 and 0 respectively, so $M_{out}(8) = 1$. MTF is normally plotted, as shown in Fig. 5.14, against a linear or logarithmic spatial frequency scale.

In the above discussion we equated line pairs density with spatial frequency. This is a convenient approximation because, as we saw in Figs. 4.10 and 4.11, a line pairs pattern is described by a primary sinusoidal spatial frequency and its odd harmonics. The amplitudes of the harmonics are significantly lower than the amplitude of the primary frequency. For the estimation of MTF we need only consider the primary frequency.

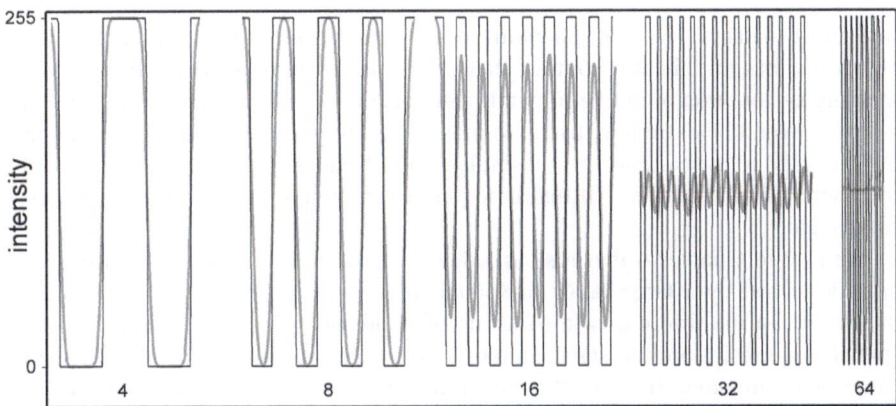

**Fig. 5.13** Intensity profiles of the original (*fine black line*) and blurred (*bold gray line*) line pairs patterns from Fig. 5.11. In this example $MTF(f)$ simplifies to the ratio of amplitudes because $I_{max} + I_{min}$ is the same for both $M_{in}$ and $M_{out}$

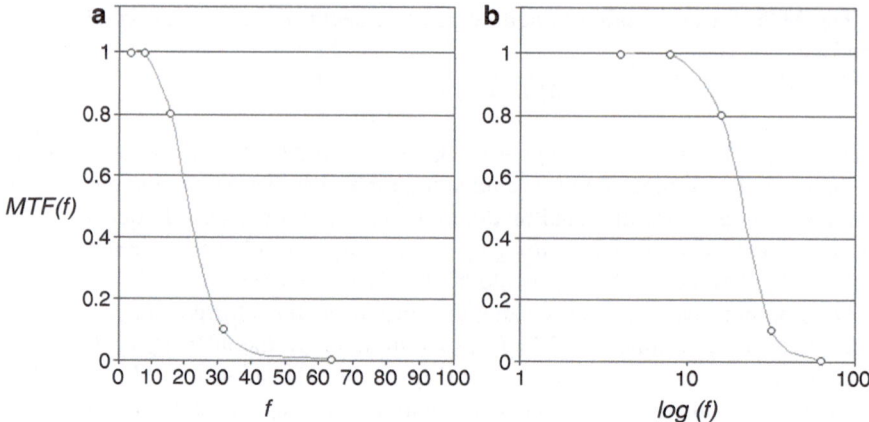

**Fig. 5.14** Linear (**a**) and logarithmic (**b**) MTF curves that describe the blurring and loss of contrast in Fig. 5.11b

Recall from Chapter 4 (Fig. 4.27) that the roughly rectangular shape of an X-ray tube's focal spot resulted in slightly different blurring in the $x$ and $y$ directions. This corresponds to different $MTF(f)$ curves for the $x$ and $y$ directions. An $MTF(f)$ curve can be used to describe the spatial resolution performance of a whole imaging system, or specific components of an imaging system. Since all MTFs have a maximum value of 1.0 and a minimum of zero, the system MTF is simply the product of the MTFs of its individual components. Analysis of the system components separately enables the identification of the most likely targets for design improvements.

### 5.5.3   The Edge, Line, and Point Spread Functions

Instead of using a line pairs phantom to test an imaging system we could use an object that has an inherent contrast that varies sinusoidally in space. For an X-ray imaging system this might be a metal plate machined to a sinusoidal profile. Such a device would be extremely tricky to make, and not particularly useful, as it could only be used to measure MTF at a single spatial frequency. In fact we don't need a line pairs phantom either. All that is required is a high contrast object with a very sharp edge.

We saw in Chapter 4 that a sharp edge or narrow line feature in an image are both described by a large range of spatial frequencies in the Fourier spectrum of the image. In the limiting case of a line of width one pixel the amplitudes of all spatial frequencies are the same. In a blurred image the amplitudes of high spatial frequencies are small or zero. If we examine the Fourier spectrum of the image of an object known to have abrupt contrast changes, such as sharp edges or narrow lines, then we should be able to determine the degree to which the amplitudes of higher spatial frequencies are attenuated by the imaging system – in other words, $MTF(f)$.

$MFT(f)$ of an imaging system can be measured in any given direction by imaging the straight edge of a suitable test object, or a 'line' object. For an X-ray system we could use the edge of a metal sheet or a thin straight wire. The orientation of the edge or line should be perpendicular to the MTF direction of interest (remember that straight image edges produce a linear Fourier spectrum feature perpendicular to the edge). The imaging system will produce an image of the edge or line which is blurred to an extent specific to the particular imaging parameters and equipment used (Fig. 5.15a, b). A plot of the intensity profile perpendicular to the edge or line shows the spatial blurring in the direction of interest – the *Edge Spread Function* (ESF) or *Line Spread Function* (LSF) (Fig. 5.15c, d). Notice that the profile of the ESF is effectively half of the symmetrical LSF.

$MTF(f)$ can be obtained by plotting the 1D Fourier spectrum of the LSF, and scaling to a maximum of 1.0 (Fig. 5.15e). In this case we use the raw frequency domain amplitudes, not their logarithms, and the zero-frequency (DC) term is first discarded because we are only interested in the amplitudes of non-zero spatial frequencies.

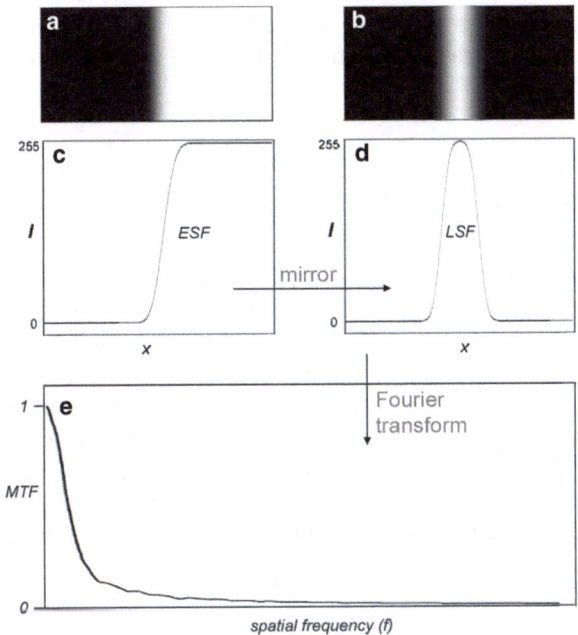

**Fig. 5.15** The edge and line spread functions can be used to determine a direction-specific MTF. (**a**) Image of a sharp high contrast straight edge. (**b**) Image of a narrow line. (**c**) The profile of the intensity in image **a** describes the *Edge Spread Function* (ESF). (**d**) The profile of the intensity in image **b** describes the *Line Spread Function* (LSF). (**e**) The MTF is the 1D Fourier spectrum of the line spread function scaled to a maximum of 1. The ESF profile needs to be 'mirrored' to produce an equivalent LSF prior to Fourier transformation

An ESF needs to be converted to an LSF before Fourier transformation to prevent formation of an edge artifact. (The 1D Fourier transform treats its input data as if it repeated infinitely in one dimension – just as the 2D Fourier transform treats an image as if it were tiled infinitely in 2D space. Notice that the right side of the ESF is 255 and the left zero so there would be a large step in the repeated profile.)

We mentioned the *Point Spread Function* PSF several times previously as the essential spatial domain description of the 2D blurring tendency of an imaging system, or a specific component of an imaging system such as the focal spot of an X-ray tube (Fig. 4.27). Obtaining a 2D MTF is simply a matter of Fourier transforming the PSF and scaling the raw 2D Fourier spectrum (minus the DC term) to a maximum value of 1.0. The value of the 2D MTF derived from the PSF is that it describes the spatial resolution of an imaging system in all spatial directions – there is no need to perform a separate measurement for each direction of interest as is required when using the line or edge spread functions. However, obtaining an image that represents the PSF is not easy for imaging systems that have an inherently weak signal such as MRI scanners. The test object used to acquire a PSF image needs to have a 'point' feature smaller than the spatial resolution of the imaging system. For plain X-ray imaging this can be a pinhole in a thin metal sheet. For CT a wire aligned with the axis of the scanner will give a PSF in the axial plane. Similarly for PET and SPECT a small bead or wire of radioactive material is used to measure either a 3D or 2D PSF respectively.

## 5.6 Contrast + Noise + Resolution

From the preceding discussion it should be clear that our ability to extract information from an image is primarily dependent on the image contrast, while the spatial scale of the information available is dependent on the spatial resolution of the image. Noise reduces our ability to detect subtle contrast differences. Noise has a much less significant effect on spatial resolution due to the tendency of our perception to perform a 2D spatial averaging of intensities. Figure 5.16 provides an illustration of the combined effects of contrast and noise on spatial resolution. This image gives an artificially strong impression of spatial resolution because our perception tends to extend the linear contrast pattern into regions that, viewed in isolation would appear to contain only noise.

## 5.7 Summary

- The primary determinants of medical image quality are *Contrast, Spatial resolution*, and *Noise*. An ideal image has high contrast, high spatial resolution, and low noise.

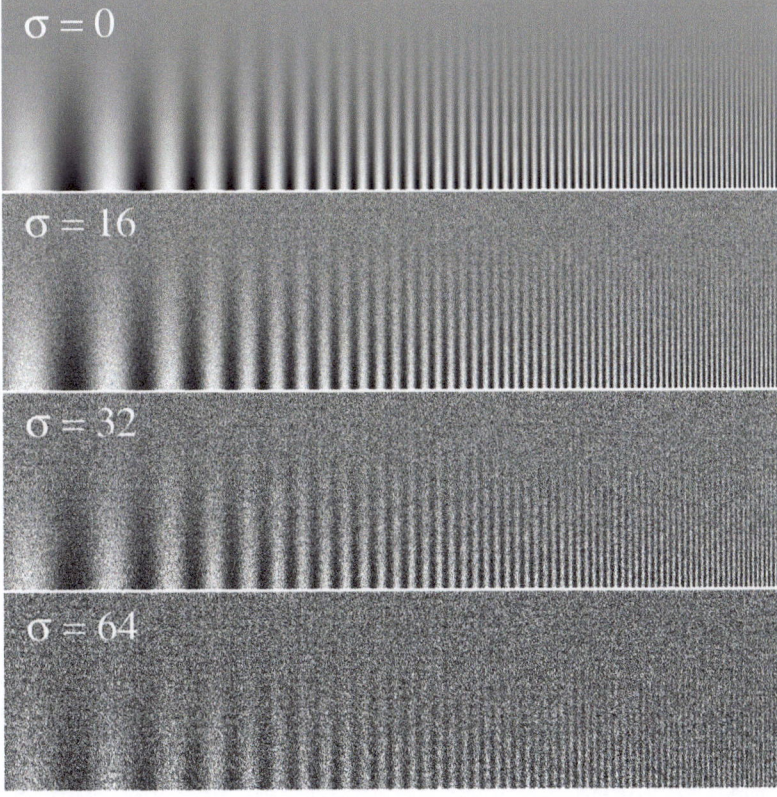

**Fig. 5.16** Illustration of the effect of noise on contrast sensitivity at different spatial frequencies. The noise PDF in this test image is Gaussian with $\sigma$ values expressed as 8-bit intensities (min = 0, max = 255). Noise decreases contrast sensitivity at all spatial frequencies but especially high frequencies

- Image contrast can be quantified as the absolute or relative difference between the average pixel intensity inside an object of interest and the average pixel intensity in the object's background.

- Most images are affected by noise – random signal variations that are superimposed on the 'true' signal of an imaged object. Noise decreases the contrast sensitivity of human perception. Low contrast objects are difficult to see in a noisy image.

- Noise can be characterized by a *probability distribution function* (PDF) – a summary of the statistical error due to noise in a recorded signal. Most image noise can be described by a *Gaussian* PDF (bell curve) with a characteristic standard deviation ($\sigma$). A Gaussian PDF means that 66% of all image pixels have an intensity error due to noise that is less than $\sigma$. A more noisy image will have a larger value of $\sigma$.

- The energy detected by an imaging system can be thought of as being either *signal* (contains information about the imaged object), or noise (contains no information and obscures the signal). The *Signal to Noise Ratio* (SNR) is a measure of the quality, or *potential* information content, of the image data. If the noise energy is random then SNR can be increased by increasing the total amount of energy measured – either by applying a more intense flow of energy, or by measurement over a longer time period.

- The *actual* information content of a medical image usually depends on *contrast* – the differences in image intensity that reveal structural or functional properties of the imaged tissue. The *Contrast to Noise Ratio* (CNR) is often a more useful measure of image quality than SNR.

- The spatial resolution of an image is dependent on the construction and geometry of the imaging system, and the interaction of the imaging energy with the imaged object. All imaging systems blur the signal from the imaged object, and the object itself may scatter some of the imaging energy. The amount of blur is characterized by the *Point Spread Function* (PSF) or the *Modulation Transfer Function* (MTF). The PSF represents the image of a single (infinitely small) point in an imaged object. The MTF is the normalized (maximum scaled to 1.0) Fourier spectrum of the PSF. The MTF describes the progressive loss of image contrast due to blurring as spatial frequency increases.

- Both PSF and MTF may be difficult to measure directly in an imaging system due to the difficulty of producing suitable test objects. A 1D approximation to the PSF can be derived from images of lines or edges which give, respectively, the *Line Spread Function* (LSF) and *Edge Spread Function* (ESF).

# Chapter 6
# Contrast Adjustment

## 6.1 Introduction

The goal of all the medical imaging methods is to record a signal with sufficient resolution in intensity, space, and possibly time, to enable a medical interpretation of tissue structure and function. The information of interest, contrast of some sort, should be recorded in the intensity data – in other words, there should be sufficient intensity resolution in the raw data to discriminate the contrast details we are interested in. For a simple X-ray image on film this would mean that we have not under or overexposed the image. If a digital image constructed from the raw data is to be interpreted visually by a human, then we want the contrast information to be easily visible. To achieve this we will usually have to adjust the contrast of the image. This can be done in general terms to ensure that the full range of raw intensity information is distributed across the full range of display intensities, and in specific terms to exaggerate the displayed intensity differences for the part of the raw data intensities that represent the tissue properties of interest.

Contrast adjustment is a *point operation*, by which we mean adjustments are made to the intensity or color properties of pixels without regard to the properties of their neighbors. In a point operation there is no change in image size or geometry. A contrast adjustment *remaps* the original pixel intensities or colors into a new range of intensities or colors using a single *mapping function* that is applied to all pixels. The mapping function is a mathematical expression that describes how each pixel's intensity is changed according to its value in the original image. When used in this context 'mapping' does not imply any spatial rearrangement of intensity information.

## 6.2 Human Visual Perception

Most medical images are interpreted visually by humans so we have to produce and process these images in a form that is optimized for human vision. The human visual system can *adapt* to light intensities ranging over an amazing ten orders of

R. Bourne, *Fundamentals of Digital Imaging in Medicine*,
DOI 10.1007/978-1-84882-087-6_6, © Springer-Verlag London Limited 2010

magnitude. However, at any instant the range of intensities (photon flux rates) that can be discriminated, that is, they are perceived somewhere on the scale between black and white, is only about four orders of magnitude (1–10,000). Within this range only about 24–30 distinct brightness *differences* can be perceived reliably. Small changes in intermediate light intensities are more easily perceived than the same size small changes in low and high intensities. This does not mean that we only ever need to display images with 24–30 distinct intensity levels. Visual perception adapts very quickly so that, as our eyes scan around an image, any local region of the image is perceived according to the local range of brightness. If we want to see fine details in an image we have to get very close to it – this improves our perception of both spatial *and* intensity details.

As well as intensity resolution we need to consider how the spatial resolution of human vision affects the images we create and view. In the region of highest visual acuity the eye has about 400 sensor cells per millimeter of retina surface. When an image is viewed with the eyes about 40 cm from the image surface each centimetre of image is focused onto about 0.4 mm of retina, or about $400 \times 0.4 = 160$ sensor cells. We would thus need an image resolution of at least 160 pixels/cm (or 400 DPI) to ensure that individual image pixels are not visually obvious. In reality the individual sensor cells in the retina are not fully independent (there are nerve cells interconnecting adjacent sensor cells) and the actual resolution limit of human vision is about 120 pixels/cm (300 DPI) at normal 'close up' picture viewing distances.

The spatial resolution of human vision can be expressed independently of viewing distance in terms of 'cycles per degree'. In this case 'cycles' refers to spatial frequency and 'degrees' to an angle in the visual field. The maximum detectable spatial frequency is about 60 cycles/degree, while the maximum *sensitivity* occurs at about 3 cycles/degree (Fig. 5.3).

## 6.3   Histograms

A simple way to assess the overall contrast of an image, or more specifically, the range of intensities in the image matrix, is to plot a histogram of the intensities (Fig. 6.1). The histogram is a plot of the number of pixels (*y*-axis) of each intensity value (*x*-axis). The *y*-axis is often labeled 'frequency' in histogram plots. This use of the term frequency refers to the frequency of occurrence of an intensity value in the image. It is NOT in any way related to *spatial frequency* and the two usages should not be confused.

In Fig. 6.1, an MRI image of a lemon, the histogram shows a very large number of black (intensity = 0) and very dark pixels corresponding to the background, which is air surrounding the lemon, and the pith which contains very little water. The MRI signal comes mainly from water, so only system noise is present in the background of the image. There are much smaller numbers of gray and white pixels representing the juicy parts of the lemon in which the water gives a relatively intense MR signal.

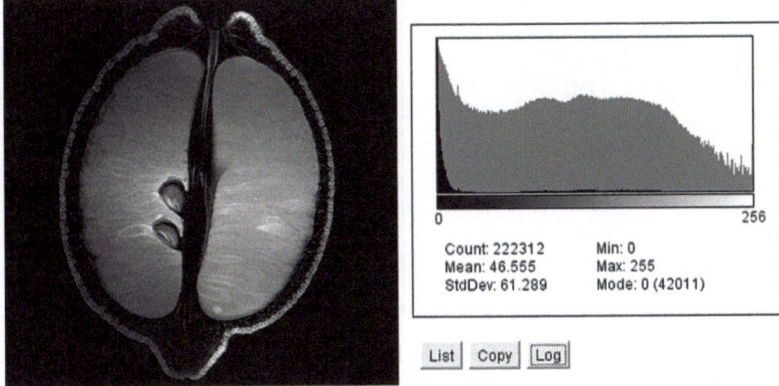

**Fig. 6.1** An MR image of an intact lemon and its corresponding histogram. The histogram is plotted on both linear (*black*) and logarithmic (*gray*) scales. The *y*-axis represents the frequency of occurrence of individual pixel intensities. Note the relatively small number of very bright pixels in the image and the corresponding low pixel count at the right end of the histogram

Because a large count of some intensities may make it difficult to assess the count of other less numerous intensity values in a histogram, it sometimes helps to plot the log of the pixel count. The log plot is shown superimposed in gray in Fig. 6.1.

In ImageJ use Menu: Analyze >Histogram to plot a histogram of the current image or current selection within an image. The 'Log' button adds in gray a supplementary plot of the log of the pixel count. Pressing 'Alt-H' on the keyboard enables plotting of a binned histogram as shown in Fig. 6.2.

A standard histogram plots one vertical bar for each possible intensity level in the image according to the bit depth of the pixel intensity data. To simplify the histogram and eliminate spikes and gaps it is sometimes useful to group intensity ranges into 'bins' and to plot one vertical bar for the total pixel count in each bin. This is termed a *binned histogram* (Fig. 6.2). Binning is also used when displaying histograms of images with a very high bit depth.

In a histogram *all* pixels with the same intensity are counted, irrespective of their position in the image, and the total count for each intensity level, or bin, is plotted as a single vertical bar. The histogram thus contains *no information* about the spatial distribution of pixels within the image. It is quite possible for two visually very different images to have identical histograms (Figs. 6.1 and 6.3, for example). Due to the lack of spatial information a histogram *cannot* be used to assess the relative contrast of objects or regions in an image. A histogram can, however, be used as part of a process that will result in the enhancement of the contrast of image objects.

A common use for histograms is to display the *dynamic range* of intensity data. In digital cameras a histogram can be used to check for *underexposure* and *overexposure* during both manual and automatic exposures. Simply reviewing a captured

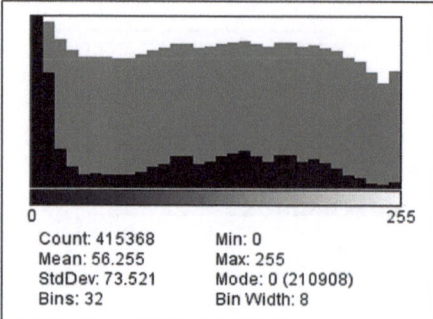 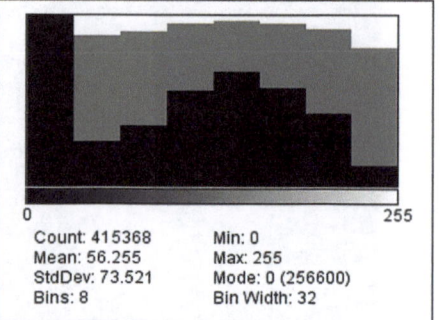

| | |
|---|---|
| Count: 415368 | Min: 0 |
| Mean: 56.255 | Max: 255 |
| StdDev: 73.521 | Mode: 0 (210908) |
| Bins: 32 | Bin Width: 8 |

| | |
|---|---|
| Count: 415368 | Min: 0 |
| Mean: 56.255 | Max: 255 |
| StdDev: 73.521 | Mode: 0 (256600) |
| Bins: 8 | Bin Width: 32 |

**Fig. 6.2** These are are 'binned' versions of the histogram of the lemon MR image (Fig. 6.1). The user specifies the total number of bins. The 'Bin Width' refers to the number of adjacent intensity levels that are combined into each binned count (Bin width = total number of possible intensity levels divided by number of bins)

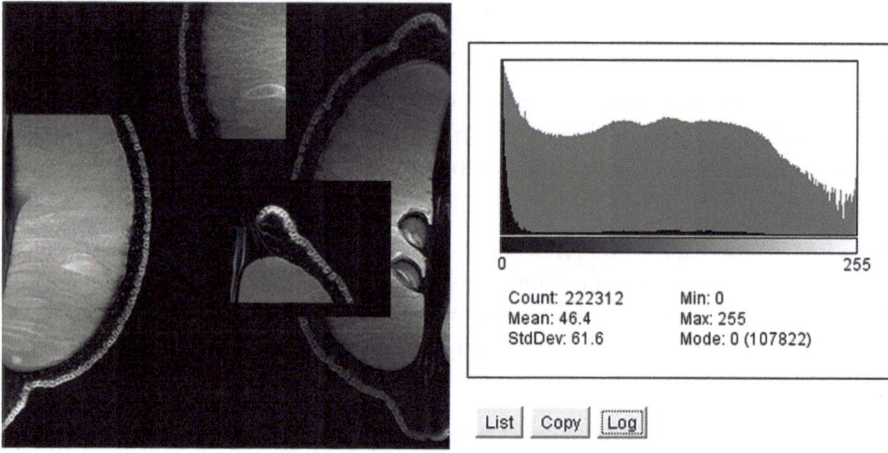

| | |
|---|---|
| Count: 222312 | Min: 0 |
| Mean: 46.4 | Max: 255 |
| StdDev: 61.6 | Mode: 0 (107822) |

List  Copy  Log

**Fig. 6.3** The lemon MR image (Fig. 6.1) sliced up and rearranged. Note that the spatial rearrangement of the intensity data has no effect on the image histogram because the histogram simply displays the number of pixels of each intensity *irrespective* of pixel position. The histogram thus contains *no spatial information* and tells us nothing about the visibility of objects in the image

image on a camera's LCD display is a very unreliable way to check exposure due to the limited spatial and brightness resolution of this display. The light metering system of a digital camera measures the average light intensity incident on the sensor, or, in more sophisticated cameras, particular regions of the sensor. For automatic exposures the camera uses these light intensity measurements to set exposure parameters (aperture, shutter speed, and sensor signal amplification) to obtain image data that are calculated to utilize the full dynamic range of the sensor and the internal image processor. However, this approach will fail if there is a very bright region in the frame that was not measured, or if the majority of the measured region is dark and a small part is very bright. In this case the image data for the bright region will be overexposed – all the pixels in the region will have the maximum possible intensity and any spatial variations in intensity within the bright region will be lost.

Just as failure to consider bright regions of the visual field can result in overexposure, the opposite problem, underexposure, can occur when exposure calculations include irrelevant bright regions. In this case subtle intensity variations in a predominantly dark subject will be compressed into a fraction of the available dynamic range and the signal-to-noise ratio will be reduced.

In manual exposure adjustment the photographer can use the histogram to check for appropriate utilization of the dynamic range *according to the particular subject and the information of interest within that subject.* This is an important point in imaging in general. We are nearly always making subjective judgements and predictions about the relevant information within the field we sample (the Field of View of a medical imaging system, or the frame of a camera exposure) and within the acquired image data. We have to make sure that the imaging device captures the information of actual or potential interest and, once captured or recorded, we have to extract that information from the image data.

## 6.4   Manual Contrast Adjustment

### 6.4.1   Contrast Stretching

Figure 6.4 shows a low contrast version of Fig. 6.1. This image is instantly recognizable as being of lower quality than Fig. 6.1. It is much more difficult to see the structural detail in the tissue of the fruit. The histogram, and our eyes, tell us that the darkest areas of the image are not very black, and the lightest areas not very white. The *dynamic range* of the data, 54–192 according to the Histogram tool, is much less than the dynamic range of the system which is 0–255. Notice that the overall profile of the histogram is the same as in Fig. 6.1, i.e. the relative numbers of dark and bright pixels is unchanged. We can predict that more information about the tissue structure is available than we can actually perceive in this image. This information may be perceptible if we can *expand* the current range of displayed intensities.

We can improve the contrast of Fig. 6.4 by expanding the dynamic range of its gray scale to the full dynamic range of the display system using the ImageJ Brightness/Contrast tool. This tool displays a miniature version of the histogram inside a box that contains a diagonal line (Fig. 6.5). Notice the changed position of this diagonal in Fig. 6.5b. The diagonal line represents a *brightness mapping function.*

A mapping function describes how the range of pixel intensities in the input image data is 'remapped' to a new range of intensities in the output image. The simplest mapping functions are *linear*. Figure 6.6 is an example of a linear mapping function that expands the central part of the range of possible input intensities to

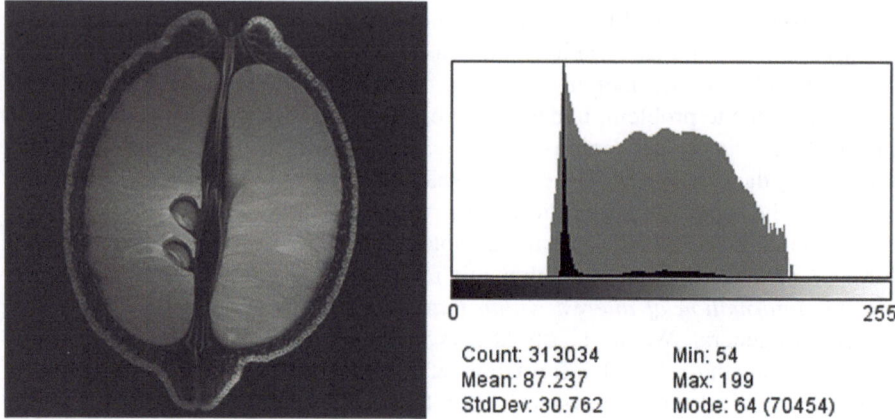

Count: 313034          Min: 54
Mean: 87.237           Max: 199
StdDev: 30.762         Mode: 64 (70454)

**Fig. 6.4** A low contrast image and corresponding histogram. The histogram indicates the absence of both high and low intensities in the image data – only about 60% of the full *dynamic range* of the display system is utilized

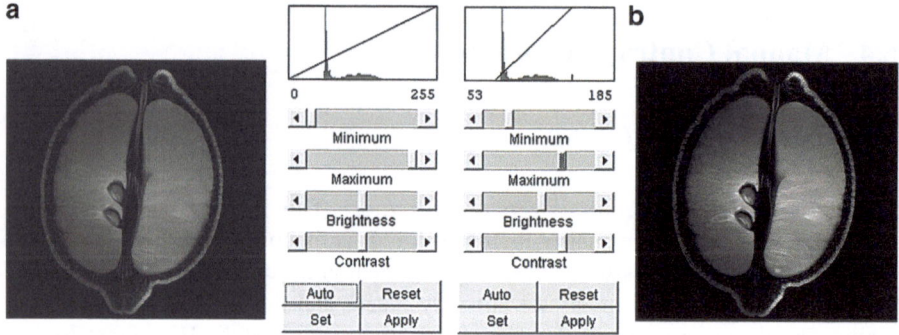

**Fig. 6.5** Adjustment of image contrast using the ImageJ tool Menu: Image>Adjust >Brightness/Contrast. (**a**) Original low contrast image. (**b**) Improved contrast resulting from adjustment of the Minimum and Maximum sliders in the Brightness/Contrast tool. Note that the contrast adjustment tool window for image **b** shows the 'input' histogram *prior* to adjustment

fill all of the possible output intensity range. In this example only a fraction of the input dynamic range is being utilized and it is this range that we expand. It would be quite possible to adjust the mapping function so that a broader range of input intensities was used (by decreasing the slope of the mapping function) with the result, for this example, that the output dynamic range would not be fully utilized and the output image contrast would be sub-optimal. Alternatively, by making the slope of the mapping function steeper, only a fraction of the actual range of input intensities would be expanded. Input intensities below the minimum of the mapping function diagonal line would all be mapped to zero (= black) and any above the top of the diagonal would be mapped to 255 (= white). The mapping of the extremities of the input value range to the minimum and maximum possible output intensities is

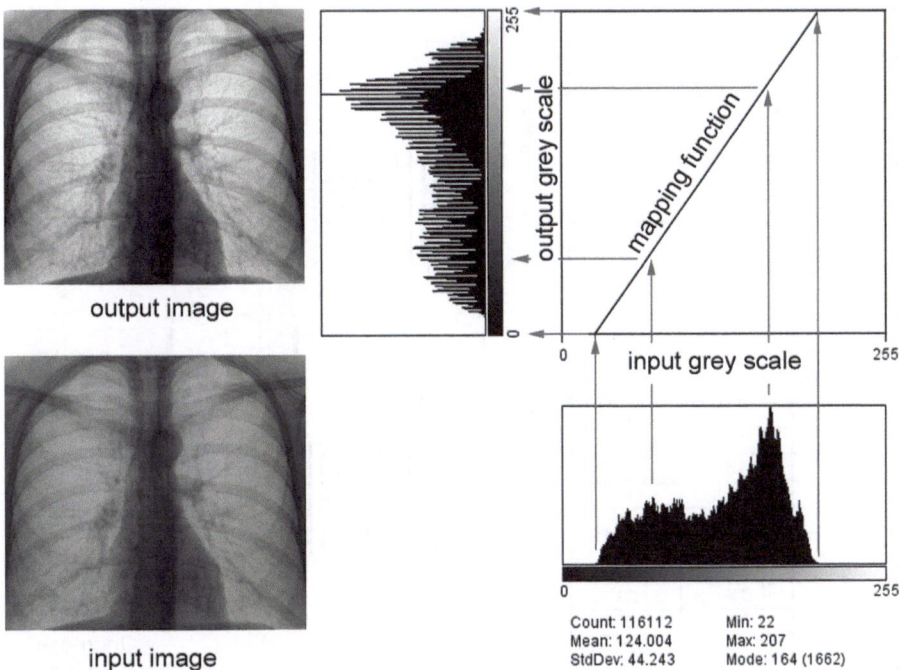

**Fig. 6.6** Adjustment of image brightness and contrast by remapping of intensity levels. In this example a *linear mapping function* is used. A selected range of intensities (22–207) has been expanded to fill the available dynamic range (0–255). The histograms are shown as summaries of the range of input and output intensities – they are not essential to the process of contrast adjustment

termed *saturation*. It is often performed intentionally as such extreme values usually represent noise rather than information.

Expanding the range of image intensities is referred to as *histogram stretching* but this term is misleading. The mapping function is NOT applied to the input histogram. It is applied to the input image data of which the input histogram is just a summary. Histogram stretching is thus a *result* of the remapping of the intensity range. It is not an essential part of the process. We can perform contrast expansions using mapping functions without ever calculating or viewing input or output histograms. We might perhaps use the input image histogram as an aid to choose the minimum and maximum values for the mapping function and this is the reason a miniature input image histogram is displayed within the ImageJ 'Brightness/Contrast' tool.

A linear contrast expansion will typically give rise to a histogram that contains regularly spaced gaps – intensities for which the pixel count is zero. Conversely, a linear contrast decrease causes regularly spaced spikes to appear – intensities with unusually high pixel counts. Both these phenomena are the result of *quantization* of the intensity values. Because adjusted intensity values must be rounded before they are stored as integers, specific integer intensity values may end up with extra or zero

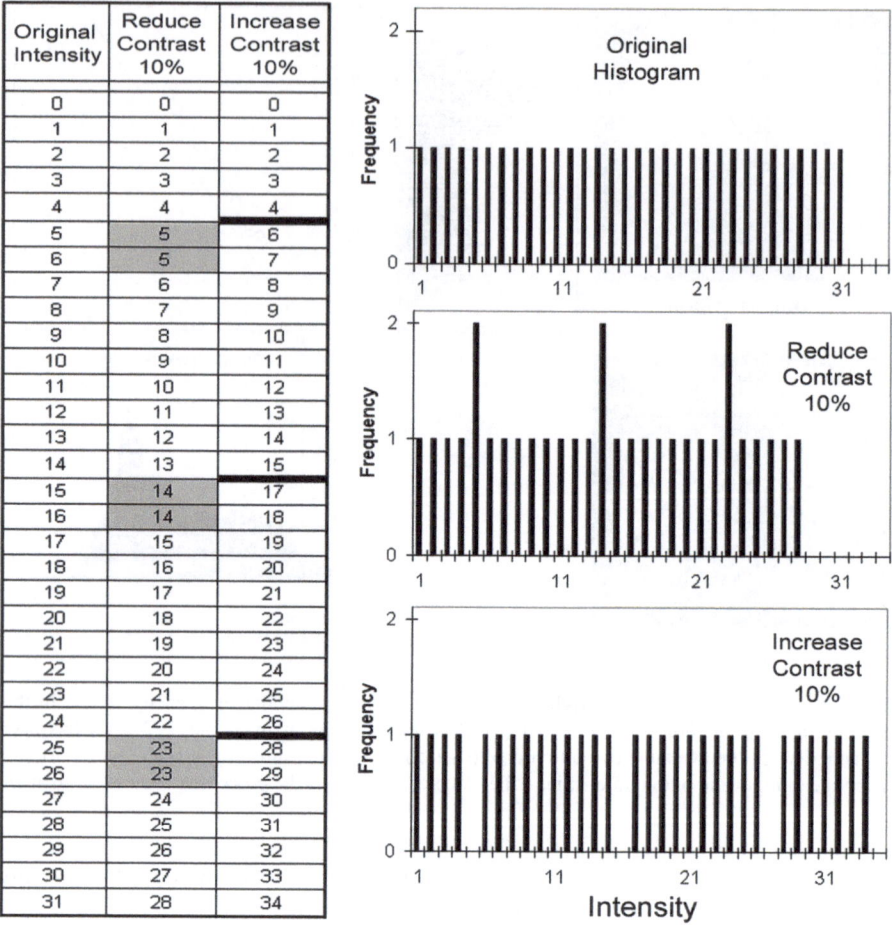

| Original Intensity | Reduce Contrast 10% | Increase Contrast 10% |
|---|---|---|
| 0 | 0 | 0 |
| 1 | 1 | 1 |
| 2 | 2 | 2 |
| 3 | 3 | 3 |
| 4 | 4 | 4 |
| 5 | 5 | 6 |
| 6 | 5 | 7 |
| 7 | 6 | 8 |
| 8 | 7 | 9 |
| 9 | 8 | 10 |
| 10 | 9 | 11 |
| 11 | 10 | 12 |
| 12 | 11 | 13 |
| 13 | 12 | 14 |
| 14 | 13 | 15 |
| 15 | 14 | 17 |
| 16 | 14 | 18 |
| 17 | 15 | 19 |
| 18 | 16 | 20 |
| 19 | 17 | 21 |
| 20 | 18 | 22 |
| 21 | 19 | 23 |
| 22 | 20 | 24 |
| 23 | 21 | 25 |
| 24 | 22 | 26 |
| 25 | 23 | 28 |
| 26 | 23 | 29 |
| 27 | 24 | 30 |
| 28 | 25 | 31 |
| 29 | 26 | 32 |
| 30 | 27 | 33 |
| 31 | 28 | 34 |

**Fig. 6.7** The origin of spikes and gaps in histograms following contrast reduction and expansion. In this simplified example the original intensities (0–31, bit depth = 5) are either reduced or increased by 10% – a linear contrast adjustment. The adjusted intensity values must be rounded to the nearest integer as only the integers 0–31 can be stored in this example. After a 10% contrast decrease rounding results in the doubling of the number of pixels with intensities 5, 14, and 23. These intensities appear as spikes in the new histogram. Conversely, after a 10% contrast increase rounding results in the absence of pixels with intensities 5, 16, and 27. These intensities appear as gaps in the new histogram. Non-linear mapping functions may lead to spikes and gaps in different regions of the histogram of the adjusted image (e.g. Fig. 6.10)

pixel counts (Fig. 6.7). The presence of spikes or gaps, or both, in a histogram is an indication that the image has previously undergone some post-acquisition contrast adjustment.

What is the difference between a mapping function and a lookup table? Both describe the range of output intensities that should be substituted for input image data. In general a mapping function is a mathematical expression used to *permanently*

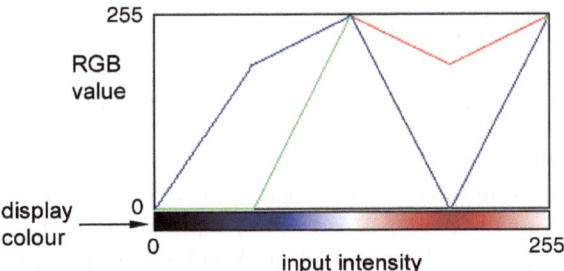

**Fig. 6.8** The ImageJ 'Show LUT' (Show Lookup Table) command displays a color plot of the RGB values in the lookup table data. This is NOT a conventional mapping function display. This example illustrates the data from the 'Union Jack' lookup table that was applied to produce Fig. 2.9c. Note that both *white* (255) and *mid-gray* (127) inputs produce a white output from this lookup table – thus some of the original intensity information is lost in a display that uses this lookup table

change the intensity data in an image matrix or image file. In contrast, a lookup table is a table that describes how to *display* image data. Application of a lookup table, or changing the lookup table of an image file, does not change the stored intensity data.

In ImageJ the Menu: Image > Color > Show LUT command (Show Lookup Table) displays a color plot of the RGB values in the lookup table data (Fig. 6.8). The display is very similar in appearance to a mapping function because the lookup table is 'mapping' each original intensity to a specific display color. The 'List' button displays the actual lookup table.

As mentioned above, a simple expansion of the histogram to match the dynamic range of the display system might improve overall contrast but may not provide optimal enhancement of the information of interest. Remember that while the histogram gives a useful summary of overall image contrast it does not tell us much about the ease of interpretation of the image. To optimize the contrast of some specific image feature (e.g. anatomy) it is generally necessary to adjust the contrast while looking at the image rather than the histogram.

In ImageJ use Menu: Image > Adjust > Brightness/Contrast to perform a linear adjustment of image contrast. This tool provides four controls to adjust only two parameters. It is easiest to make adjustments using either just the top two sliders, or just the bottom two.

## 6.4.2   Window and Level

In medical imaging, especially CT, it is common to refer to linear contrast adjustments with the terms *Window* and *Level* (Fig. 2.4). In this context the *Window* (or Window Width) describes the width of the range of image data intensities that will be expanded to fill the dynamic range of the display. This corresponds to the horizontal 'span' of the diagonal part of the linear mapping function. The *Level* (Window Level) specifies the center of the Window range. All pixels outside the window are saturated – changed to either black or white. Figure 6.9 illustrates the use of the ImageJ 'Window&Level' tool to selectively enhance the contrast of either bone or metal implants in a projection X-ray image.

In CT the image data represents calculated CT numbers rather than arbitrary intensity units. Since the CT number is directly related to the linear attenuation coefficient of the imaged tissue, defined widows and levels are used to optimize the display of contrast for particular tissue types. Examples of bone and soft tissue windows were illustrated in Fig. 2.5.

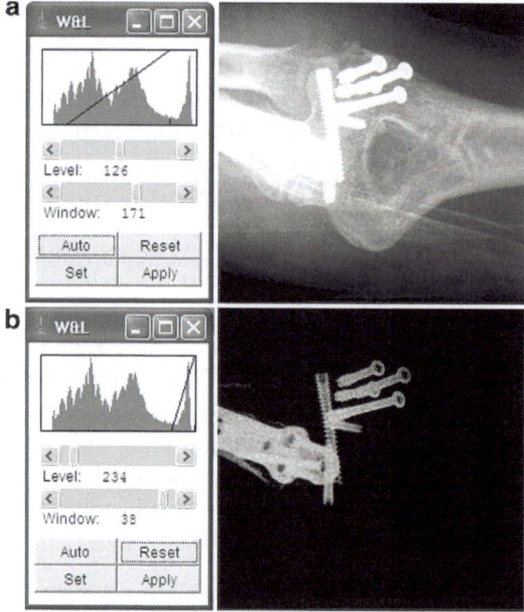

**Fig. 6.9** Selective linear contrast adjustment using Window and Level. The *Window* describes the width of the range of image data intensities that will be expanded to fill the dynamic range of the display. The *Level* specifies the center of the Window range (see also Fig. 2.5). In this example window **a** is selected to optimize the contrast for bone and soft tissue. Window **b** eliminates all tissue and highlights the detail of the surgical metal implants

In ImageJ the Menu: Image > Adjust > Window/Level tool behaves identically to the Contrast and Brightness slider controls in the 'Brightness/Contrast' tool.

### 6.4.3 Nonlinear Mapping Functions

A linear contrast expansion is ideal for distributing the intensity data of interest across the full dynamic range of a display system without changing relative intensities within that range. However, we may also want to enhance the contrast of images (or regions of images) which already utilize the full dynamic range available. We want to enhance differences in intensity without saturation of the brightest or darkest intensities. Put another way, we want to make the dark regions darker, and the light regions lighter without losing intensity information due to saturation. This contrast adjustment can be performed with a non-linear mapping function (Fig. 6.10).

## 6.5 Automatic Contrast Adjustment

Why bother with a histogram at all? So far we have used histograms for nothing more sophisticated than finding the minimum and maximum intensities in an image and checking that the acquired data lies appropriately within the dynamic range of the system. The histogram is much more useful in automated contrast adjustment.

### 6.5.1 Normalization

The contrast stretching operation just described using the Brightness/Contrast or Window/Level tools can also be performed automatically if we specify the type of adjustment we want to make. When a linear mapping function is used to stretch the dynamic range of the input image to fill the available dynamic range of the storage or display system the operation is called *Normalization* (or, somewhat inappropriately, 'Histogram Normalization').

The simplest form of normalization stretches the histogram as we did above to set the input minimum intensity to zero in the output and the input maximum to the available maximum (255 if bit depth is 8). The problem with this method is that in a predominantly low contrast gray image a single pixel with a very high or low value will prevent the full enhancement of the image. We can get around this problem by assuming that the information of interest does not include extreme intensities and

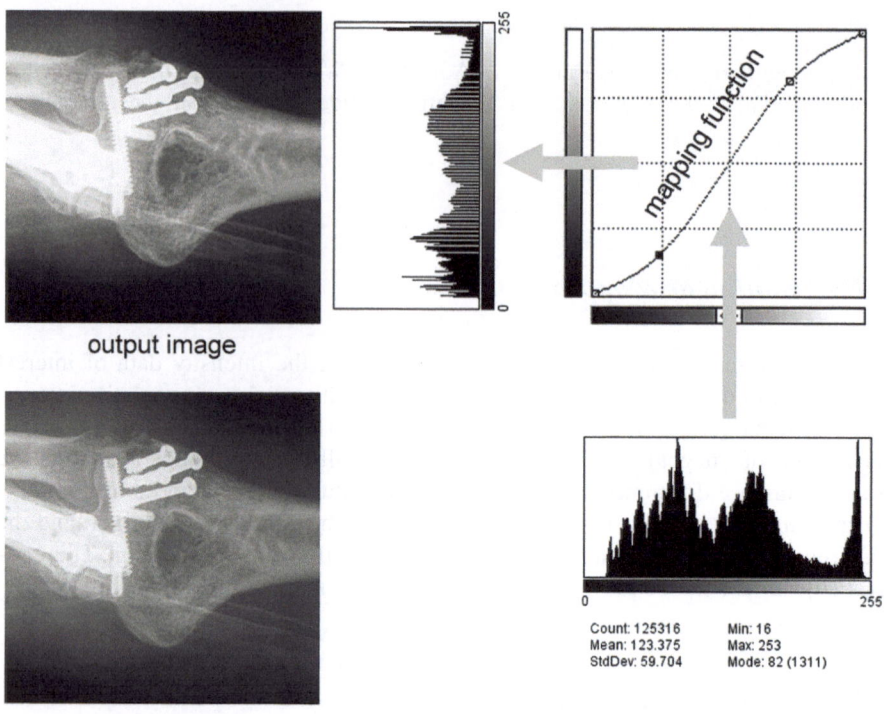

output image

input image

**Fig. 6.10** A non-linear mapping function can be used to enhance contrast without saturation of the highest or lowest intensities. In this example a sigmoid shaped mapping function is used to darken all low intensities and brighten all high intensities. The effect is to enhance the visibility of bone details without losing all soft tissue detail. Note the presence of gaps in the center of the output image histogram, due to contrast expansion of the middle range of intensities, and spikes in the histogram at low and high intensities, due to contrast compression of these intensities (described in Fig. 6.7). The illustrated mapping function is from Adobe Photoshop. Non-linear manual contrast adjustment is not currently available in ImageJ

permit a small percentage of pixel values (the extreme ones) to saturate. The linear contrast stretch will then be performed over a range slightly less than the full input range.

In ImageJ use Menu: Process > Enhance Contrast to perform automatic normalization with adjustable saturation.

Figures 6.11 and 6.12 illustrate the use of normalization to improve contrast in an MR image. The percentage of total pixels (dark and bright) deemed to be extreme has been set in the 'Saturated Pixels' box of the ImageJ 'Enhance Contrast' tool. As the Saturated Pixels setting is increased the peak of the original histogram

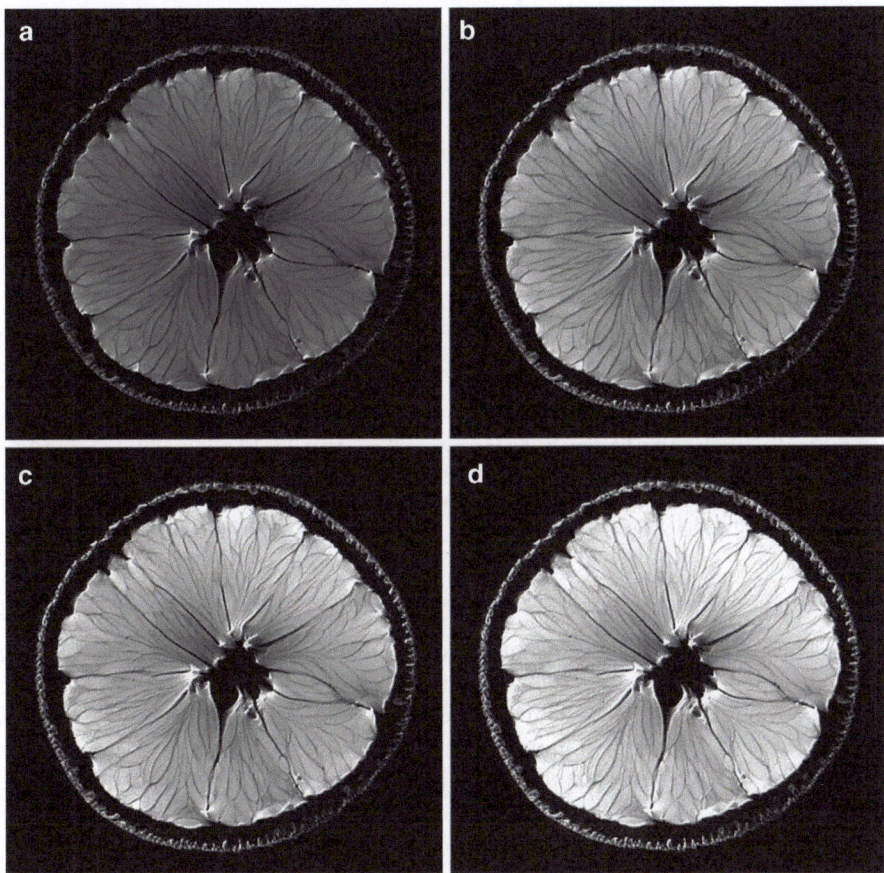

**Fig. 6.11** Image normalization with variable saturation. (**a**) Original low contrast MR image of an orange. (**b**) Normalized image with Saturated Pixels = 1%. (**c**) 2%. (**d**) 8%. The percentage of total pixels, dark and bright together, was set in the 'Saturated Pixels' box of the ImageJ 'Enhance Contrast' tool. Note that in the extreme case, when the percentage of pixels excluded from the normalization is too large, image detail is lost in the brightest regions. The histograms of these four images are shown in Fig. 6.12

broadens and shifts towards the high intensity end of the scale, consistent with the overall increase in brightness of the image. Note that in the extreme case, when the percentage of pixels excluded from the normalization is too large, image detail is lost in the brightest regions.

## 6.5.2 Histogram Equalization

If we temporarily ignore the fact that histograms tell us nothing about the spatial distribution of intensity information in an image and think about images *only* in terms

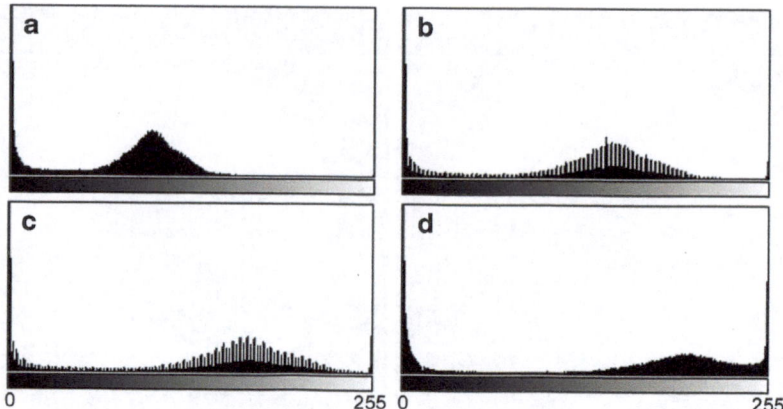

**Fig. 6.12** Histogram normalization using increasing 'Saturated Pixels' settings (0, 1, 2, and 8%). The histograms are those of the orange MR images shown in Fig. 6.11. As the 'Saturated Pixels' setting is increased the central peak of the original histogram broadens and shifts towards the high intensity end of the scale, consistent with the overall increase in brightness of the image

of their histograms, then we might reasonably assume that an image with 'ideal' contrast properties would have its intensity information evenly distributed across the available range. Such an image would have a histogram with a flat profile – there would be roughly the same count of pixels for each possible intensity. *Histogram Equalization* is an automated process that does just this – it attempts to redistribute an image's pixel intensities to produce a histogram with equal counts for all possible intensities.

From the regular histogram we can plot a *cumulative histogram* which is a running sum of the total pixel count starting from the lowest possible intensities and progressing to the highest (Fig. 6.13). The cumulative histogram thus *never* decreases as we scan from left to right – its slope is always positive or zero. When an image contains no pixels in a particular intensity range – a gap in the regular histogram – the cumulative histogram will be flat in that range. When there are a lot of pixels in a particular intensity range – a peak in the regular histogram – the cumulative histogram will increase rapidly.

You may notice from Fig. 6.13 that the plot of the cumulative histogram looks similar to a mapping function. What would happen if we used the shape of the cumulative histogram as a mapping function to adjust the contrast of an image? This is exactly how equalization is implemented! The process is illustrated in Fig. 6.14. Notice, however, the raw histogram of the equalized image is not flat, as we should expect from equalization, but contains many spikes and gaps. This is because we are dealing with quantized intensity data. The spikes and gaps occur for the same reason we saw them after linear contrast adjustments (Fig. 6.7). To see the expected flat histogram we need to bin the intensities into ranges. The use of the cumulative histogram to describe a mapping function that will flatten an image's histogram is more obviously demonstrated with a simple image such as that shown in Fig. 6.15.

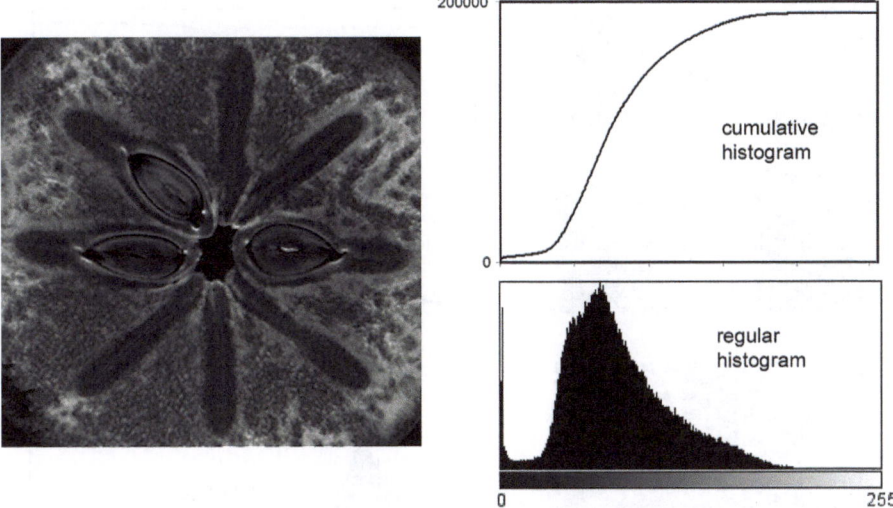

**Fig. 6.13** The cumulative histogram represents a 'running sum' of the pixel counts in the regular histogram or, equivalently, the total number of pixels with an intensity less than or equal to a specific intensity. Notice that the cumulative histogram is flat in intensity ranges where there are low or zero pixel counts and increases rapidly in ranges that have high pixel counts

Because equalization is directly based on an image's histogram, and the spatial information in the image is absent from the histogram, the visual effects of equalization depend very strongly on the individual image. For most images it turns out that equalization using a mapping function based on the simple cumulative histogram produces an image that appears, to a human, to be too bright and lacking in overall contrast detail. To reduce this common adverse result a mapping function based on the *square root* of the pixel counts can be used (Fig. 6.16). This is the default method in ImageJ. Figure 6.17 compares the effects of normalization and equalization applied to a low contrast X-ray image.

Equalization is an option in the ImageJ 'Enhance Contrast' tool. The default algorithm uses the square root of the pixel count from the original histogram to define the cumulative histogram and the mapping function. Hold down the 'Alt' key when pressing the 'OK' button to perform equalization using the raw pixel counts. If a ROI is defined within the image when the Enhane Contrast tool is opened then the option 'Update All When Normalizing' will be available. This option uses only the data inside the ROI for calculation of the mapping function, but applies the mapping function to *all* pixels in the image. In the case of equalization the ROI-based mapping function is always applied to the whole image.

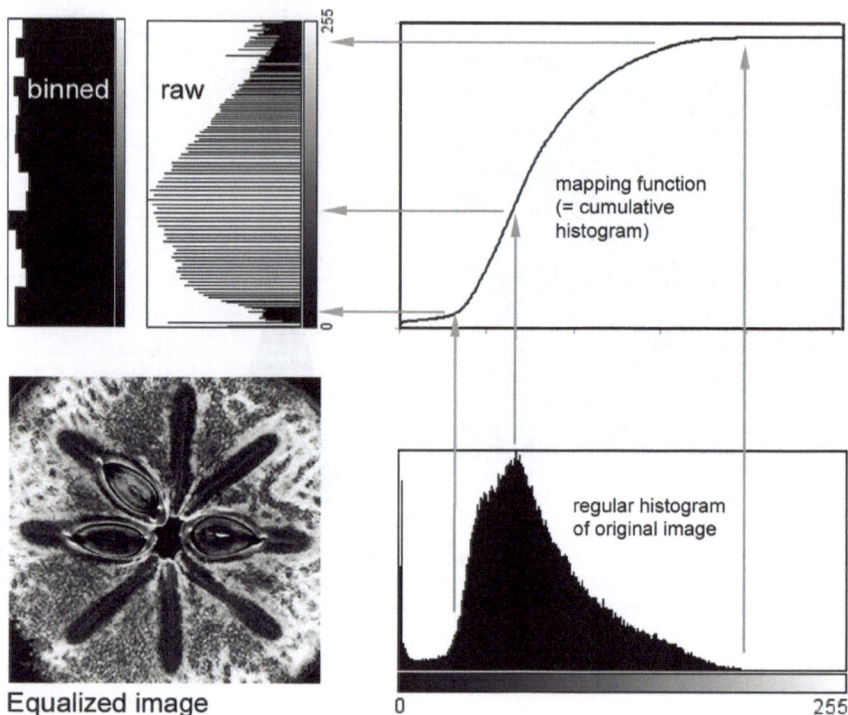

**Fig. 6.14** Equalization of the low contrast MR image shown in Fig. 6.13. The equalization process uses the cumulative histogram to define the shape of the mapping function. The histogram of the equalized image has roughly the same number of pixels in all intensity ranges, however, due to quantization of the intensity data, the expected flat histogram is usually only seen after binning of intensities

### 6.5.3 Histogram Specification

We saw above that for many images equalization, or flattening, of the regular histogram does not produce visually ideal contrast properties. Most 'good looking' images actually have a roughly bell-shaped regular histogram. This is also the case for medical images. If we assume that some 'ideal' histogram shape exists, at least for a particular type of image, then perhaps we could adjust suboptimal images by making their histograms conform to this specific ideal. This can be done with a modification of the equalization process called *Histogram Specification*.

The equalization method uses the cumulative histogram of an image as the form of a mapping function that, in principle, flattens the regular histogram. We can, at least in theory, reverse this process – that is, take an equalized image with a flat histogram and use the mapping function backwards to get back to an image with a histogram the same as the original image before equalization. The 'reverse' mapping function is the inverse of the equalization mapping function. Now consider our image that has an 'ideal' histogram. If we were to equalize this image we would use

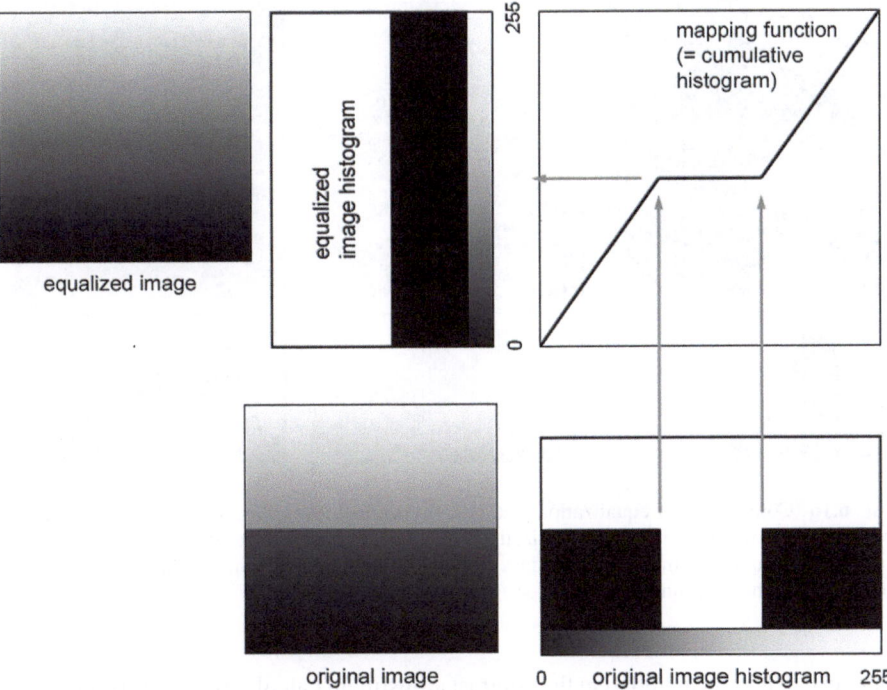

**Fig. 6.15** A simplified demonstration of the equalization process. The cumulative histogram of the original image describes the form of a mapping function which will flatten the original histogram. The original image here has dark and light intensities but no pixels in the middle intensity range. Its regular histogram thus has a large central gap and its cumulative histogram a corresponding central flat region. The effect of using the cumulative histogram as a mapping function is to stretch the low and high intensity ranges such that they now meet in the middle and the gap in the original histogram is eliminated. The equalized image has an equal number of pixels at each possible intensity level

a mapping function that flattens its histogram. The inverse of this mapping function, let's call it the 'specification function', will change *any* image that has a flat histogram into one that has the 'ideal' histogram. Histogram specification is simply the two step process of first equalizing an image and then applying a specification function to 'unflatten' the histogram and make it the desired shape.

Histogram specification is a good idea in principle but suffers from the intensity quantization problem and only really works for images of quite similar spatial composition.

## 6.5.4 Region-Specific Contrast Adjustments

Automated normalization and equalization often fail to produce improvements in image contrast because irrelevant image information, for example background noise

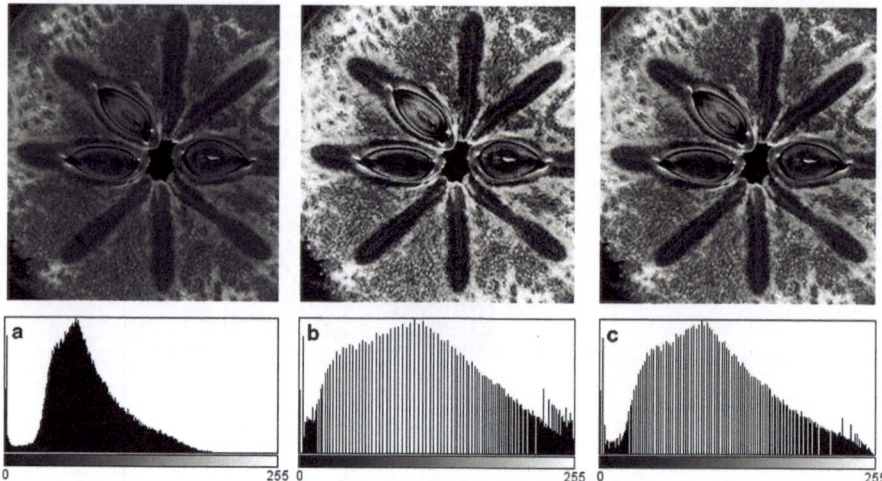

**Fig. 6.16** Comparison of equalization methods. (**a**) original low contrast image. (**b**) Equalization based on the raw pixel count. (**c**) Equalization based on the square root of the pixel count. For most images the square root method (**c**) produces a more visually attractive image, however, notice that the details of the persimmon seeds are easier to see in image **b**

or overlaid text, is included in the contrast adjustment calculations. This problem can often be avoided by defining within the image a Region Of Interest (ROI) on which to base the calculations. The mapping functions thus generated can be applied just to the ROI or to the whole image (see Section 6.6 for an example).

### 6.5.5 Binary Contrast Enhancement – Thresholding

Sometimes the only information we are interested in is the boundaries between regions in an image. We might for example want to select all the regions in an X-ray image that represent bone, or all the white matter from an MR image of the brain. The general name for this process – dividing an image into distinct regions – is *Segmentation*. Under ideal conditions distinct anatomical regions will be associated with unique ranges of intensities. Better still, we might be able to say that there is a single cutoff intensity above or below which all pixels belong to a particular category, bone or not-bone, say in our X-ray image. In this ideal case we could define two distinct image regions simply by changing the intensity of any pixel that excedes the threshold value to 255, or white, and changing any pixel less than or equal to the threshold to zero, or black. We would thus create a *binary image*, or *mask*, that contains only two possible pixel intensities. Binary images are typically shown in pure black and white, with the choice of which region is shown black and which white being somewhat arbitrary.

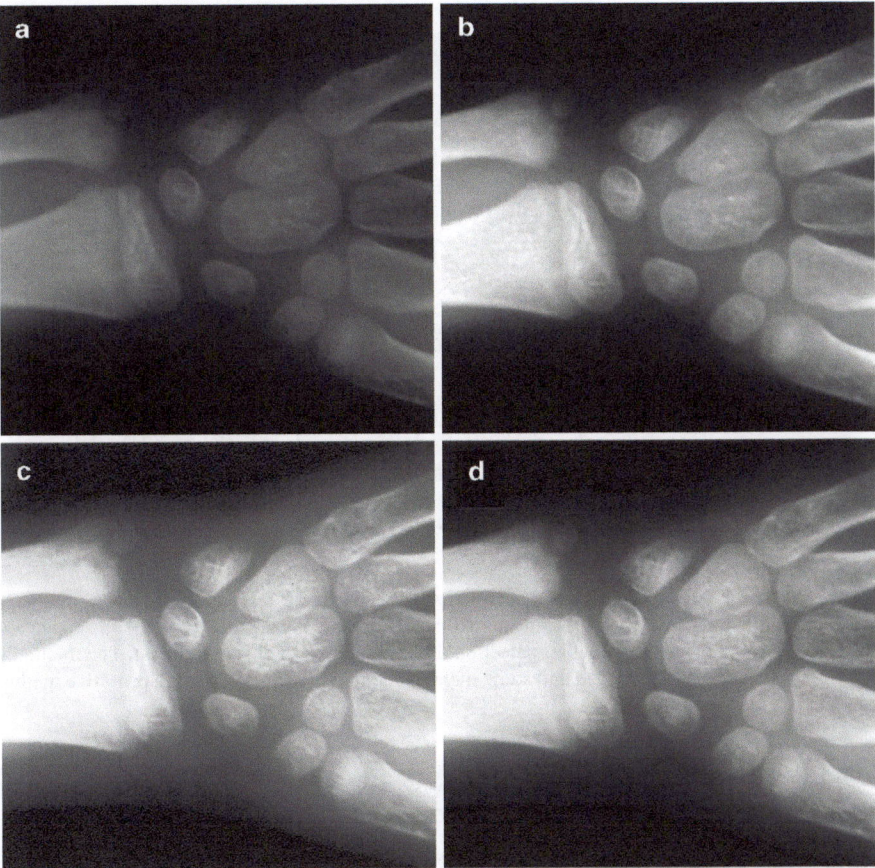

**Fig. 6.17** Comparison of effects of normalization and equalization on contrast of an X-ray image. (**a**) Original image. (**b**) Normalized image (0% saturated pixels). (**c**) Equalization based on raw pixel counts. (**d**) Equalization based on square root of pixel counts. Equalization based on raw pixel counts overbrightens the image and exaggerates background noise. For this particular image equalization based on the square root of pixel counts enhances contrast detail in most of the anatomy, without too much noise exaggeration, however, some detail in the largest bone is lost

Imagine we are only interested in the bone details of an X-ray image such as Fig. 6.18. We can see that the bones are generally bright and select an appropriate threshold intensity (135 in this case) to create a binary 'bone' mask. A possible problem with this binary image is that it includes implant metal in the bone mask. Since most of the metal appears brighter than bone in the original image use of a *second* upper threshold enables the exclusion of most of the metal while retaining most of the bone.

When storing binary images we can save a lot of computer space by using the values zero and one, instead of zero and 255, to store the intensity information. For an eight bit image this represents an 88% saving since the required bit depth

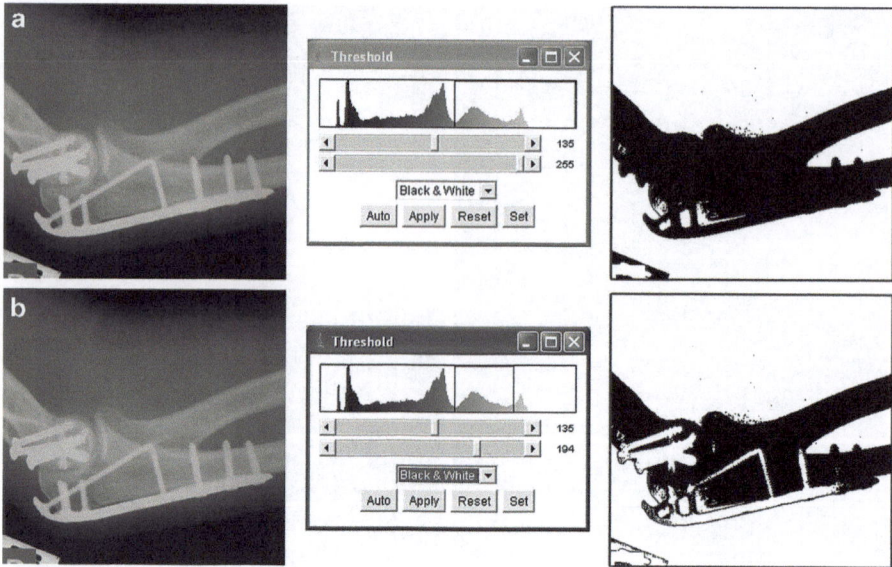

**Fig. 6.18** Use of thresholding as a crude selection of the bone regions of an X-ray image. A single threshold intensity (**a**) results in inclusion of implant metal with bone. Using upper and lower threshold intensities (**b**) enables selection of bone without inclusion of metal. The method is imperfect – some bone is excluded and some metal is included. Thresholding is a primitive method of image segmentation

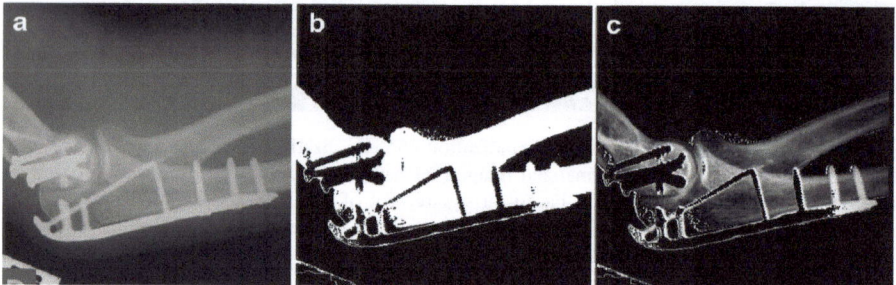

**Fig. 6.19** A binary mask created by thresholding can be used to exclude irrelevant information from an image. Here the 'bone' mask from Fig. 6.18b is inverted and used to remove soft tissue and metal from the original image. Each original pixel intensity is multiplied by the corresponding mask pixel value (0 or 1). Image c = a × b

is only one. Obviously a great deal of image information is lost in the thresholding process but this is quite intentional. In thresholding an image we are only interested in the question: 'Which of two categories does a pixel belong to?'. Used as a mask a binary image can be multiplied pixel-wise by the original image to remove all the image information that is of no interest. For example, Fig. 6.19 illustrates the use of our 'bone only' mask (after inversion) to remove the soft tissue and metal implant details.

The ImageJ Menu: Image > Adjust > Threshold tool creates a binary image based on the selection of upper and lower thresholds. All pixel intensities above the upper threshold, or below the lower, are changed to the maximum possible (white). Pixels between the thresholds are changed to zero (black). This tool also enables a color display of the binary image. The tool Menu: Process > Image Calculator can be used to apply a mask to another image.

## 6.5.6 Hardware Contrast

Our discussions about image contrast adjustment made the implicit assumption that all image viewing devices will produce identical displays from identical image data. This assumption is rarely justified for gray scale images and is generally not true at all for color images. In medical imaging it is essential to check and calibrate image display devices to achieve consistent contrast visibility.

### 6.5.6.1 Gamma Adjustment

For a display monitor the output light intensity ($I$) is related to the source voltage ($V$) by a power law:

$$I \propto V^{\gamma} \tag{6.1}$$

Monitor gamma ($\gamma$) describes the adjustment that is made to image intensity data to achieve a *perceived* display intensity that is linearly related to the image data intensity. Figure 6.20 illustrates the effect of gamma adjustment. When $\gamma = 1$ no change is made to the raw intensity data with the result that the gray values appear too light. In Fig. 6.20 the gamma values of 1 or 1.2 produce a roughly linear gray scale gradient *in the printed image*. On a CRT monitor display, which is based on light emission rather than light reflection as is the case for this printed image, a gamma in the range 2.2–2.5 is usually appropriate. The light output of LCD monitors cannot be described by a simple power law as for CRT monitors. Proper adjustment of the output of LCDs requires calibration with a photometer and software that can modify the graphics card lookup table.

Some computer operating systems include a gamma adjustment tool. This tool is used to make small adjustments to gamma in order to make a patch of mid gray appear to have, when viewed at a distance, the same brightness as a grid of closely spaced black and white lines (Fig. 6.21). This adjustment sets the midpoint on the gamma curve to enable a roughly linear gray scale display but does not account for contrast differences due to luminance or overall brightness.

Image *acquisition* devices also have non-linear responses to input energy intensity that can be approximated by a power law. The important message here is that the

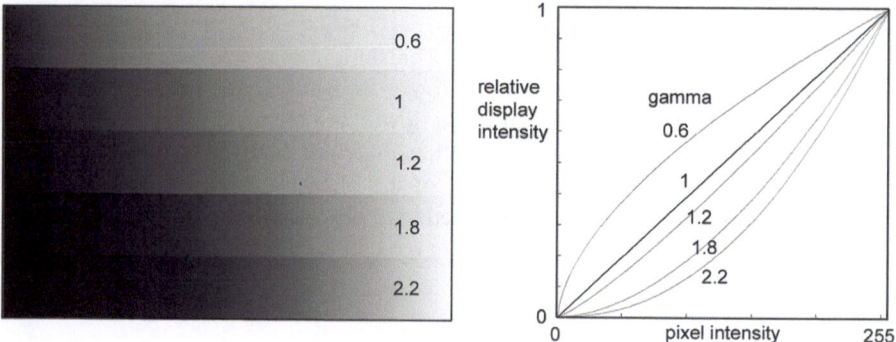

**Fig. 6.20** Illustration of the effect of gamma adjustment where display intensity depends on input signal according to a power law: $I \propto V^{\gamma}$. When $\gamma = 1$ no change is made to the raw intensity data with the result that the gray values appear too light. Here the gamma values of 1 or 1.2 produce a roughly linear gray scale gradient *in the printed image*

**Fig. 6.21** Part of the *Adobe Gamma* CRT monitor adjustment tool. The aim is to adjust gamma until the black/white bar pattern appears to have the same brightness as the central mid-gray region. This adjustment sets the midpoint on the gamma curve to enable a roughly linear gray scale display

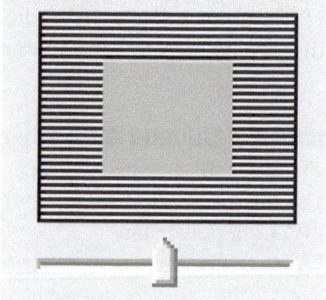

signal intensities that are recorded as pixel values in an image file will, in general, NOT be linearly related to the energy emitted by the imaged objects, nor to the luminance of a display (or density of a print) created from the image file.

### 6.5.6.2 The DICOM Gray Scale Display Function

Gamma adjustment should not be confused with *monitor calibration* which is necessary to achieve consistent and predictable *perception* of image contrast differences. Monitor calibration requires a colorimeter device which is periodically placed on the face of the monitor to check and adjust calibration. Some of these devices can also automatically adjust the display intensity to compensate for changing room lighting conditions.

Medical image displays should be calibrated according to Part 14 of the DICOM Standard (www.nema.org) which defines a *Grayscale Standard Display Function* (GSDF). The purpose of the GSDF is to ensure that diverse medical image display devices produce images of comparable quality in terms of speed and accuracy of diagnostic interpretation. Some newer medical monitors have built-in automatic GSDF calibration using an internal photometer.

## 6.6 Practical Example. Adjusting the Contrast of a Magnetic Resonance Microimage

Finally, let's take a look at a practical example of a contrast adjustment problem. In Fig. 6.22 we see a newly acquired MR image of an excised mouse brain immersed in formaldehyde solution in a cylindrical glass bottle. The image was acquired in an 11 Tesla microimaging system so the spatial resolution is much higher than could be achieved in a typical 1.5T or 3T human MR scanner. Nevertheless, this image illustrates a number of the problems commonly seen in clinical medical images. Most of the image background, aside from the text, is black – there is no MR signal detected outside the glass bottle because there is no water (more specifically, no hydrogen nuclei) outside the bottle. For the same reason we see no signal from the glass walls of the bottle. The detected signal arises from water protons in the brain tissue and in the formaldehyde solution outside the brain. The information of interest – the structure of the brain – is revealed by subtle contrast differences within the brain tissue, and the overall shape of the brain by the contrast difference between the brain tissue and the formaldehyde solution. The tissue structure information is squeezed into a range of gray scale intensities running from mid-gray to near white.

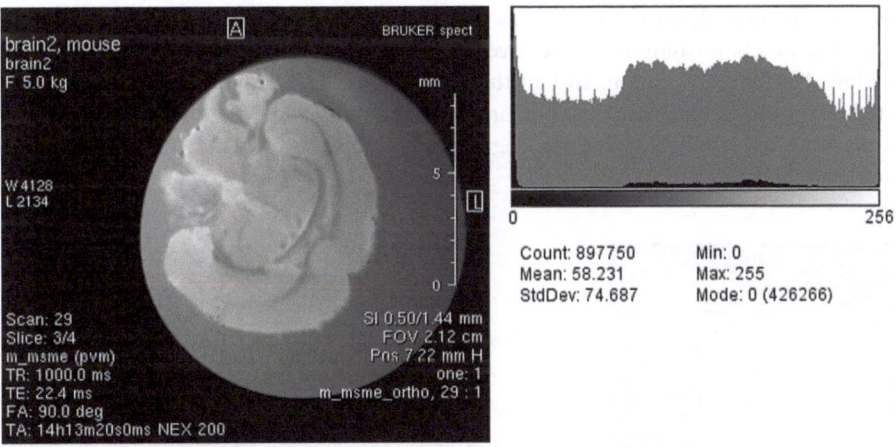

**Fig. 6.22** High-resolution MR image of a mouse brain as typically displayed on the console of the imaging system, in this case a Bruker 11 Tesla system. The image uses the full dynamic range of the display system but within the brain tissue the contrast is relatively poor (image courtesy of Professor Bill Price, University of Western Sydney)

The scanner software overlays the image with text describing significant acquisition parameters. This text contains the only truly white pixels in the image.

Looking at the histogram we see a large Uluru-like hump in the middle range of intensities. It is tempting to assume that this hump represents the brain tissue, but such an assumption might be a mistake. Remember that the histogram is simply a display of the relative number of pixels of each intensity – it tells us nothing about the spatial distribution of those pixels.

In order to improve our ability to see the structure of the brain we need to adjust the contrast of the image. In particular, we want to *expand* the contrast of the region of interest, the brain tissue, and we don't particularly care about what happens to the contrast outside the brain tissue, except perhaps we don't want to lose the readability of the text. We can make a manual contrast adjustment with the ImageJ Brightness/Contrast tool, as shown in Fig. 6.23. We increase the Minimum slider to make the darkest parts of the brain almost black, and decrease the maximum slider to make the brightest parts of the brain almost white. It is important that we make these adjustments while looking at the image and not just the histogram. We want to enhance the contrast without losing structural information. If, for example, we set the Maximum slider a little too low, some of the anatomical detail in the brightest regions of the brain is lost.

We have just discussed a method of making a *linear* adjustment to the image contrast based on our visual assessment of changes in the region of interest as we manually adjust the contrast – a linear remapping of the grayscale. In some images we could achieve the same improvement in contrast, or possibly improve on it, if we used an automated contrast adjustment method such as histogram equalization. Figure 6.24 illustrates the output image after we equalize *the whole image*. The result is unsatisfactory in terms of visualization of the brain structure because the equalization algorithm was based on the cumulative histogram of the whole image. The pixels forming the white overlaid text and the dark background have been included in the calculation. In fact the noise in the background has been exaggerated. It would be preferable to perform the equalization based on a histogram that represents only the pixels in the region of interest. We can do this by manually defining the ROI.

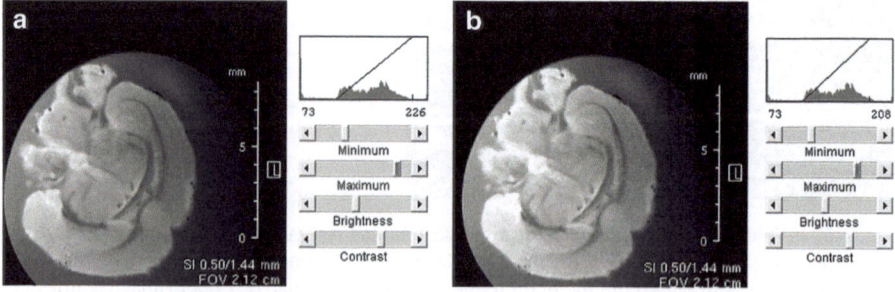

**Fig. 6.23** Linear contrast enhancement of the brain tissue in Fig. 6.22 using the ImageJ 'Brightness/Contrast' tool. In image **b** the 'Maximum' slider has been set too low. Note the loss of tissue structure details in the bright regions. In image **a** the enhancement is close to ideal

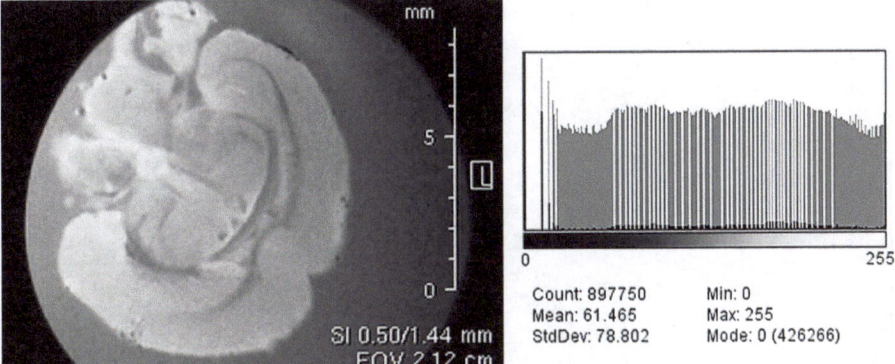

**Fig. 6.24** Contrast adjustment based on equalization of the histogram of the whole image, including text and background. Note the noise now visible in the background and the lack of contrast enhancement in the brain tissue

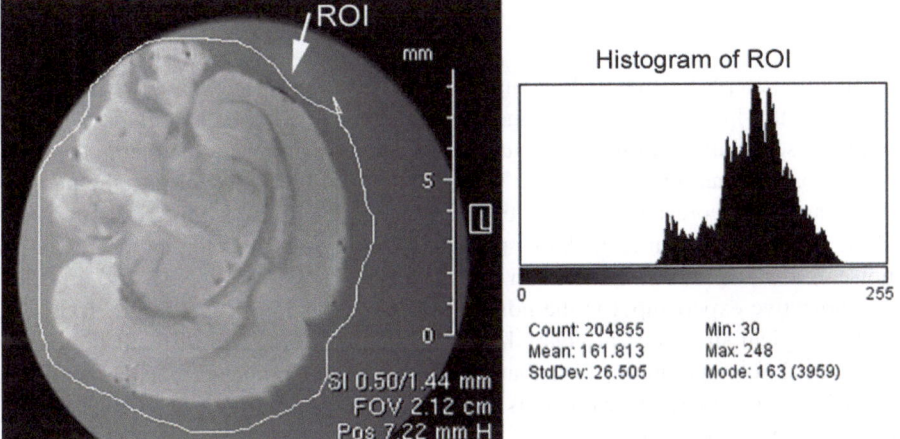

**Fig. 6.25** Manual selection of a Region Of Interest (ROI), and display of the histogram of the ROI, permits exclusion of irrelevant pixel intensities from the contrast adjustment process

ImageJ provides several tools for selection of specific regions of an image on which subsequent operations or calculations will be performed. In Fig. 6.25 we have used the freehand selection tool to define a ROI that just includes the mouse brain. When a ROI is currently defined within an image the Histogram tool displays only the data from within the ROI. Note the similarity between the shape of this histogram and the central hump of the histogram of the whole image (Fig. 6.22). In this particular image our earlier warning about making assumptions about the spatial origin of a histogram feature turned out to be over-cautious. Now the mapping function for the equalization will be based on just the pixel data within the ROI. The equalization can then be applied to all the pixels in the image (Fig. 6.26).

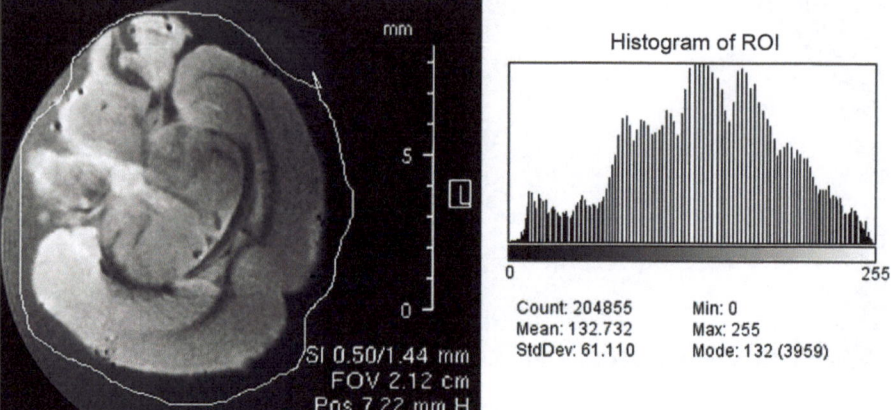

**Fig. 6.26** Contrast adjustment of the whole image based on equalization of the histogram of the ROI defined in Fig. 6.25. Now there is good tissue contrast. The loss of contrast details outside the ROI is of no consequence

In this example we manually defined the ROI. We made a subjective expert judgement about which part of the image contained the important information and we applied user input to constrain the automated calculations and adjustments. It seems trivial for a human inspector of the mouse brain MR image to point to the parts of the image that represent the brain tissue, but this task is far from trivial for a machine which has to somehow know not just what a brain looks like but also what is noise or text added after acquisition of the image.

Subjective expert input is the norm in medical imaging because human bodies, and biological structures in general, are extremely diverse and difficult to characterize in simple geometric ways that are amenable to automated machine analysis. Only in a very few specific contexts can a machine more reliably recognize pathology than an expert human.

## 6.7 Summary

- Image data is usually measured and stored with greater precision than the intensity resolution of human perception. Contrast adjustment is the modification (remapping) of stored pixel intensities in order to make specific intensity information more visually obvious in the image display.
- A histogram provides a summary of the range and frequency of occurence of pixel intensities in a digital image. However, the histogram provides *no* information about the spatial distribution of pixel intensities within the image. Two completely different images may have identical histograms.

- *Linear* contrast adjustments use a linear mapping function to expand a range of input pixel intensities. *Normalization* is the process of expanding the *whole* input intensity range to the full available dynamic range.

- A *fraction* of the input range may be expanded with a linear mapping function to optimize the display of particular image features. In CT, for example, defined *Windows* are used to display specific ranges of the stored CT number data using the full dynamic range of the display device.

- *Equalization* uses an image's *Cumulative Histogram* to describe the form of a non-linear intensity mapping function. The effect is, theoretically, to flatten the regular histogram and thus produce an image with the same number of pixels of each possible intensity. Because intensity data is quantized the equalized histogram usually has many spikes and gaps, however, a *binned histogram* of the image will be approximately flat.

- *Thresholding* is an extreme form of contrast adjustment. All pixels intensities within a specified range are remapped to the available maximum intensity, and all other pixels are remapped to the available minimum intensity. Thresholding converts a gray scale image to a pure black and white (binary) image.

- Imaging detectors typically have a non-linear response to input energy. Similarly, image display devices typically have a non-linear response to input data. The response of many devices can be described by a power law and can be adjusted by a single parameter – *gamma*.

# Chapter 7
# Image Filters

## 7.1 Introduction

Filtering is one of the most common applications of image processing. By filtering we mean the processing of the image in order to remove or reduce a particular unwanted component, for example noise, or to enhance or extract a particular set of features, such as edges. We can apply filters in both the spatial and frequency domains and many filters have equivalents in both domains. On the other hand, some filters operate exclusively in one domain or the other. Because the terminology used to describe both spatial and frequency domain filters often refers to spatial frequency features and effects we will first discuss the frequency domain filters.

## 7.2 Frequency Domain Filters

### 7.2.1 Ideal Filters

We saw in Chapter 4 how the 2D Fourier transform resolves a spatial domain image into its spatial frequency components. One of the most significant uses of this transformation is that in the frequency domain we can manipulate the spatial frequency components independently. We could, for example, remove all of the components with a spatial frequency less than some chosen cutoff. This is what we did in Fig. 4.20 of Chapter 4 and it is referred to as *high pass* filtration because the filter *passes* (does not attenuate) the spatial frequencies that are higher than the chosen cutoff.

If, instead of completely removing the low frequency components (i.e. zeroing low frequency amplitudes in the frequency domain matrix), we merely reduced their amplitudes, then the result would be a spatial domain image in which the edges, and possibly the noise, were enhanced or exaggerated. By adjustment of both the cutoff frequency and the degree of attenuation of the low frequencies we can change the amount and nature of the enhancement.

R. Bourne, *Fundamentals of Digital Imaging in Medicine*,
DOI 10.1007/978-1-84882-087-6_7, © Springer-Verlag London Limited 2010

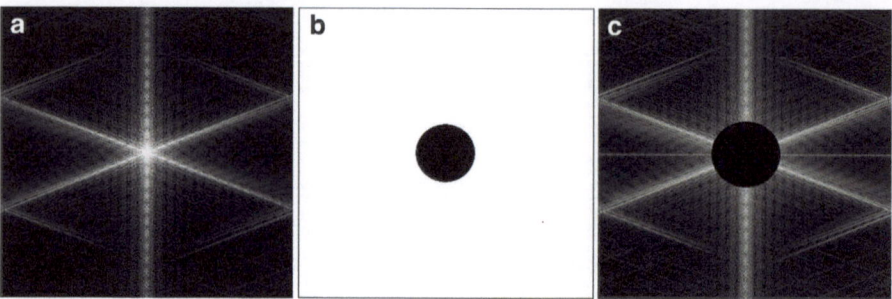

**Fig. 7.1** Application of an Ideal high pass filter. The filter mask (**b**) is applied to the frequency domain data (represented by Fourier spectrum **a**) by *element-wise* multiplication. In an Ideal filter the mask has only two possible values: 1 (*white* in this example), or 0 (*black* in this example). All amplitudes below the *cutoff* are multiplied by zero and thus reduced to zero. All amplitudes above the cutoff are multiplied by one and thus remain unchanged. Image **c** (= **b** × **a**) shows the Fourier spectrum of the filtered frequency domain data

A frequency domain filter can be thought of as a mask applied to the complex frequency domain matrix. The mask itself is a matrix of real numbers ranging from zero to one and representing the degree of attenuation of the spatial frequency amplitudes. The mask and the frequency domain matrix have the same dimensions and they are multiplied elementwise to obtain the filtered matrix.

The high pass filter mask that was applied to Fig. 4.17b to produce Fig. 4.20a is shown in Fig. 7.1. In this particular mask there are no values other than zero and one. Where the mask is white (1) the frequency domain data will be unchanged, and where the mask is black (0) the frequency domain amplitudes will be zeroed and there will be 100% attenuation. The magnitude of the inverse Fourier transform of the filtered complex data is the filtered spatial domain image (Fig. 4.26).

The filter represented by the mask in Fig. 7.1 is referred to as an *Ideal high pass* filter because of its rectangular profile. The attenuation changes from zero to 100% at the cutoff frequency and there are no partial attenuation values. The 'Ideal' part of the description is a relic of analog signal processing where a filter with a perfectly rectangular attenuation profile is desirable but impossible to construct from capacitors and inductors. The implementation of true Ideal filters has only become possible with the advent of digital signal processing. In image processing, however, the Ideal filter is, in practical terms, *not* ideal since it usually produces *ringing artifacts* which appear as a series of lines of decreasing intensity lying parallel to the edges in an image. This artifact, due to application of an Ideal high pass filter, can be seen in Fig. 7.2 which is an enlarged and contrast enhanced view of Fig. 4.20b.

Application of an Ideal high pass filter to frequency domain data may create an abrupt step-shaped change in the frequency domain amplitudes − all the amplitudes below the cutoff of the filter will be zeroed. Just as the step-shaped intensity change in the spatial domain image of Fig. 4.5 gave rise to a series of non-zero amplitudes in the frequency domain, inverse Fourier transformation of a frequency domain data set that contains a step shaped amplitude change will give rise to a regular series of

Fig. 7.2 An Ideal filter,
despite its name, typically
produces *Gibbs ringing*
artifacts in the filtered image.
In this closeup view of
Fig. 4.20b the ringing artifact
appears as faint lines parallel
to the sharp edges in the
original image

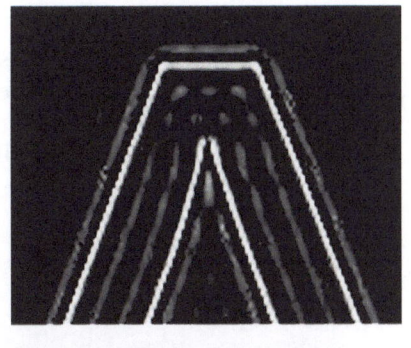

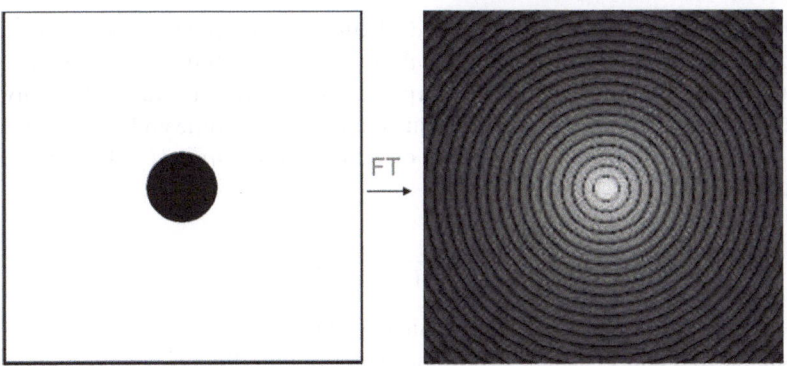

FT
→

**Fig. 7.3** The tendency of the Ideal filter to produce ringing artifacts can be predicted by examination of the Fourier spectrum of the filter mask. Although the mask is applied in the spatial frequency domain, the amplitudes of the Fourier transform of the mask (the Fourier spectrum) represent the intensity variations that will be associated with all edges when the filtered frequency domain data is converted to a spatial domain image

non-zero intensities in the spatial domain — this is the ringing artifact, also known as *Gibbs ringing*. Considering the similarity of the equations for the Fourier transform and its inverse (Chapter 4), this should not be a surprising result.

We can predict the tendency of an ideal filter to produce a ringing artifact if we look at the Fourier spectrum of the filter mask (Fig. 7.3). It comprises a series of concentric rings that diminish in intensity towards the higher spatial frequencies. Fig. 7.3 is similar to Fig. 4.8 except that the series of non-zero terms are now present in all directions radial from the center. Remember that we saw an edge feature in the spatial domain gives rise to a linear feature in the frequency domain that lies at right angles to the original edge. The circular edge of the ideal filter mask thus gives rise to a whole series of linear features which together form the rings seen in Fig. 7.3.

Ideal filters can be applied in ImageJ using Menu: Process > FFT > Custom Filter... First select a mask using Menu: Process > Filters > Show Circular Masks... and scroll through the available sizes. On finding a suitable sized mask use Menu: File > Save As > PNG to save the selected mask as an image. Open the mask image. Click on the window of the image to be filtered then open the Custom Filter tool (Menu: Process > FFT > Custom Filter...). Select the chosen mask image in the 'FFT Filter' window and press OK.

## 7.2.2 Butterworth Filters

The ringing artifact produced by an Ideal filter can be reduced or eliminated by changing the profile of the filter so that the cutoff is less abrupt. There are a number of filters that can be used for this purpose. We will discuss the commonly used Butterworth filter as it illustrates the adjustment of both profile and cutoff frequency.

The Butterworth filter has a profile, or *Transfer Function* ($H$), described by the equation:

$$H = \frac{1}{1 + (\frac{r}{R_o})^{2n}} \tag{7.1}$$

where, $H$ = filter mask magnitude (Range: 0–1),
$r$ = spatial frequency (distance from the center of the filter),
$R_o$ = nominal cutoff frequency (at which $H = 0.5$),
and $n$ = order of the filter.

The transfer function ($H$) describes how much of the original frequency domain amplitude is *transferred* to the filtered frequency domain. At the spatial frequencies where $H = 1$ there will be zero attenuation. Where $H = 0$ there will be 100% attenuation. A distinct feature of the Butterworth filter is that it has a very flat profile away from the cutoff frequency. Figures 7.4–7.6 illustrate how the filter profile changes as the order ($n$) of the filter is changed. As $n$ increases the slope of the cutoff increases. For very high $n$ the Butterworth filter will be effectively identical to an Ideal filter. Changing $n$ to a negative number inverts the filter profile, changing the filter from low pass to high pass (Fig. 7.7).

As we did with the Ideal filter we can assess the tendency of a particular Butterworth filter to produce a ringing artifact by examining its Fourier spectrum. Figure 7.6 shows the masks and corresponding Fourier spectra of Butterworth filters of order 1, 2, and 4. The mask with the steepest profile ($n = 4$) produces a mild ringing artifact which could also have been demonstrated by 1D Fourier transformation of the filter profile. The effects of the second order ($n = 2$) high pass and low pass Butterworth filters on some MRI data are illustrated in Fig. 7.8.

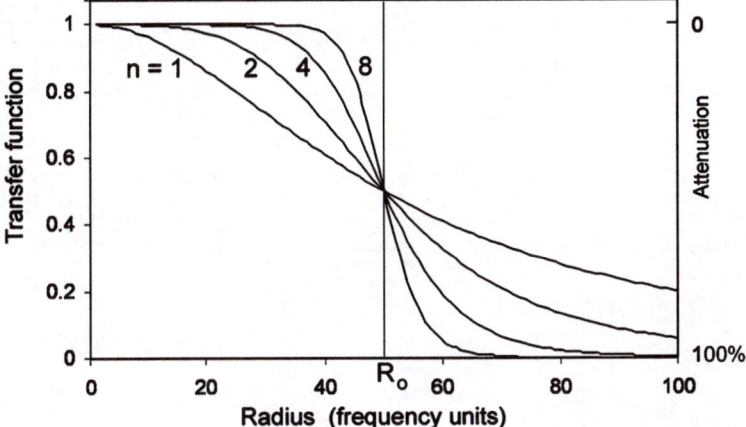

**Fig. 7.4** A *Butterworth filter* has an adjustable mask profile. The transfer function describes how much of the original frequency domain amplitude is *transferred* to the filtered frequency domain (Eq. 7.1). The *cutoff* frequency, $R_0$, is the frequency at which attenuation is 50%. Note that as the order, $n$, of the filter increases the steepness of the profile increases. Zero frequency lies at the origin of these plots, but is at the center of a 2D filter mask. Typical masks are shown in Fig. 7.6

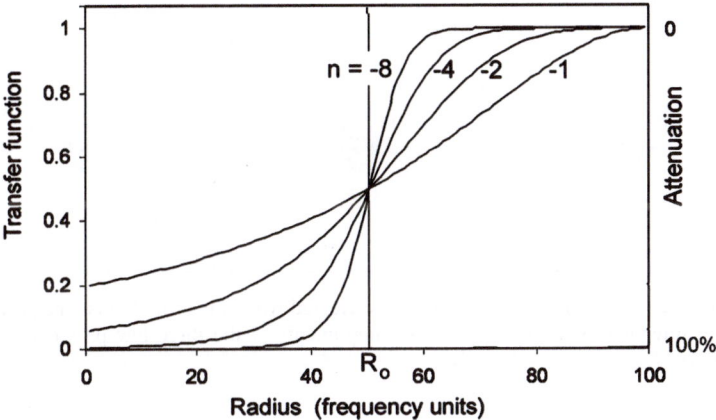

**Fig. 7.5** Butterworth high pass filter profiles. Negative values of $n$ convert a low pass filter to a high pass filter

A Butterworth filter tool is not included in the standard ImageJ package. It *may* be available as a plugin, though this seems to depend on when you search.

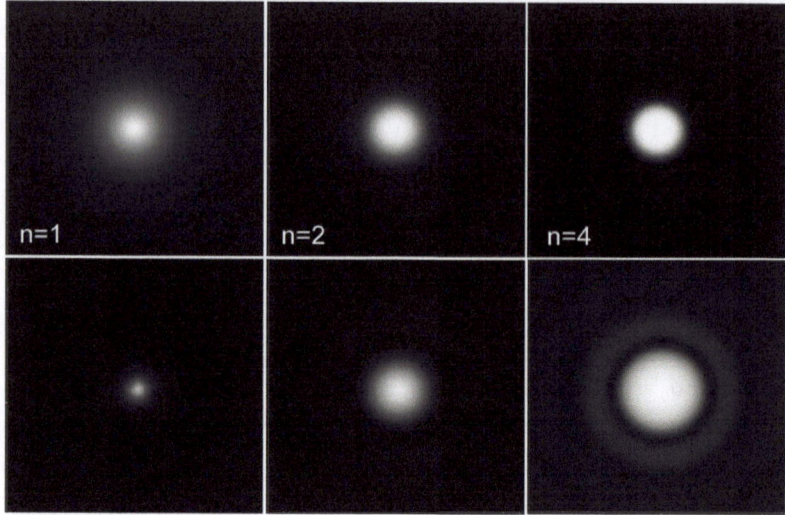

**Fig. 7.6** Butterworth low pass filter masks (*top*) and their Fourier spectra (*bottom*). Note the outer ring feature in the Fourier spectrum when $n = 4$, which indicates the filter may produce a ringing artifact along sharp edges. Zero frequency lies at the center of these masks. The profiles plotted in Figs. 7.4 and 7.5 represent the intensity changes along a radius of a mask

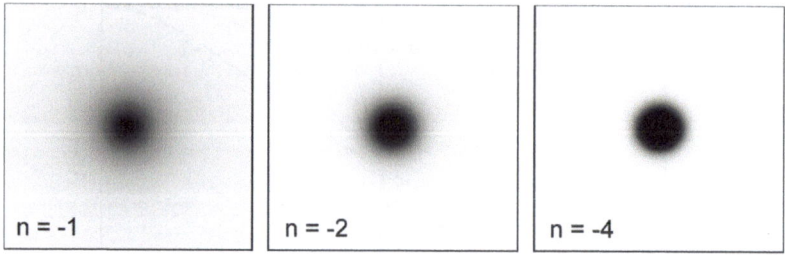

**Fig. 7.7** Butterworth high pass filter masks are created when the order (*n*) is a negative number. These three filter masks are the high pass compliments of the three low pass masks shown in Fig. 7.6 above

## 7.2.3 Gaussian Filters

Blurring, or smoothing, is a simple method of reduction of Gaussian and quantum mottle noise in images. The most commonly used blurring filters have a Gaussian profile where the transfer function ($H$) is described by an equation of the form:

$$H = e^{-\frac{r^2}{2\sigma^2}} \tag{7.2}$$

Where $H$ = filter mask magnitude (Range: 0–1),
$r$ = spatial frequency (distance from the center of the filter),
and $\sigma$ determines the cutoff frequency.

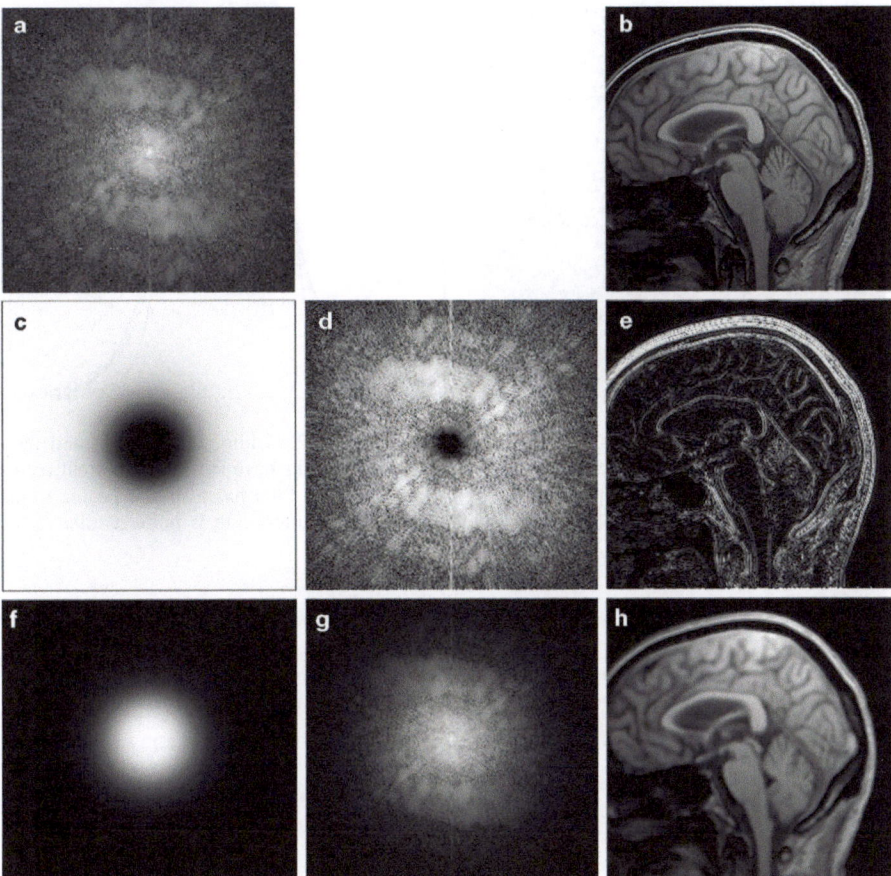

**Fig. 7.8** High pass and low pass Butterworth filters applied to MRI data. (**a**) Fourier spectrum of raw data. (**b**) Anatomical image which is the magnitude of the inverse Fourier transform of the complex data represented by **a**. (**c**) High pass filter mask. (**d**) High pass filtered Fourier spectrum. (**e**) Anatomical image created from high pass filtered raw data. (**f**) Low pass filter mask. (**g**) Low pass filtered Fourier spectrum. (**h**) Anatomical image created from low pass filtered raw data. Because MRI data is *acquired* in the spatial frequency domain extra low or high spatial frequency data can be acquired in order to emphasize either contrast or edge detail in the anatomical image

The Gaussian function also describes the probability distribution of a normal statistical distribution with standard deviation $\sigma$ (Chapter 5). In the Gaussian filter we do not have separate parameters to control the cutoff frequency and the steepness of the profile. The basic shape of the profile is constant and its width is determined by $\sigma$. As $\sigma$ is decreased the cutoff frequency decreases and the steepness of the profile increases (Fig. 7.9).

A significant feature of the Gaussian filter is its predictability. It *never* causes a ringing artifact. This could be demonstrated by examination of the Fourier spectrum of a Gaussian mask, however, this is not necessary. The Fourier transform of a Gaussian is another Gaussian. This fact makes it very easy to implement a Gaussian filter in either the frequency domain or the spatial domain.

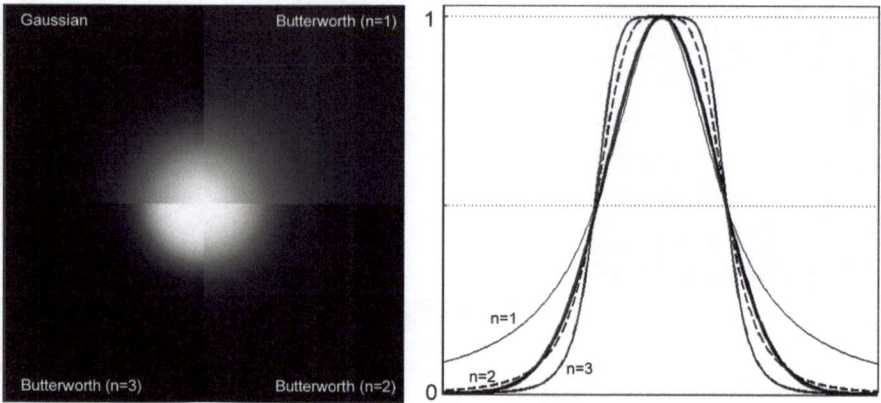

**Fig. 7.9** Comparison of Gaussian and Butterworth ($n = 1, 2, 3$) filter masks (*left*) and their profiles (*right*). The heavy trace is the Gaussian. All these masks have the same nominal cutoff frequency. Note that at low frequencies the $n = 1$ Butterworth filter has a similar profile to the Gaussian. At frequencies higher than the cutoff the $n = 2$ Butterworth is more similar to the Gaussian

In ImageJ the Gaussian blur filter is normally implemented by spatial domain convolution (Section 7.3) (Menu: Process > Filters > Gaussian Blur....). The effect is identical to application of a Gaussian mask in the frequency domain.

### 7.2.4 Band Stop Filters

One of the most impressive applications of frequency domain filtering is the removal of periodic noise — a feat which is virtually impossible to perform in the spatial domain. Figure 7.10a shows an image badly affected by periodic noise which manifests as a series of evenly spaced bright/dark bands affecting the whole image. If we focus our attention on the noise we see that it appears to have a well-defined spatial frequency — about 20 cycles per image width. We might expect that upon transformation of this image into the frequency domain the periodic noise would appear as high amplitude terms of a specific frequency, or narrow range of frequencies. This is indeed the case, as can be seen in Fig. 7.10b. In the Fourier spectrum the noise appears as a pair of white spots symmetrically displaced 20 pixels either side of the center zero frequency. If we now attenuate these high amplitude terms in the frequency domain data (Fig. 7.10c), then perform an inverse Fourier transform, we obtain Fig. 7.10d. The periodic noise is now very substantially reduced without any major detriment to the image quality.

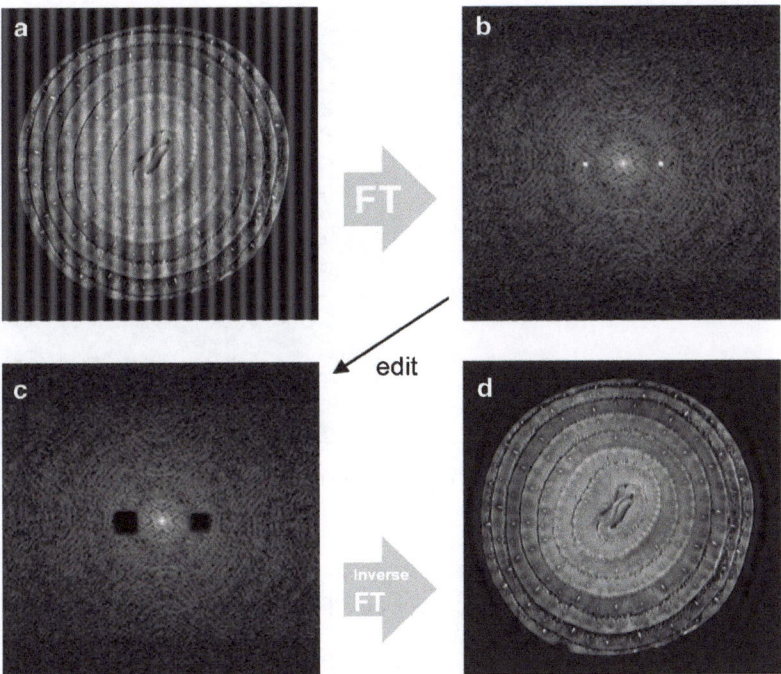

**Fig. 7.10** Removal of periodic noise by frequency domain filtering. (**a**) Image affected by periodic noise. (**b**) The Fourier spectrum of image **a** shows a pair of high intensity points on either side of the center. (**c**) The Fourier spectrum after suppression of the high amplitudes corresponding to the periodic noise. (**d**) Filtered image produced by inverse Fourier transformation of the edited frequency domain data

A more general way to achieve the same result would be to zero all the amplitudes of the small range of spatial frequencies that characterize the noise, i.e. draw an annulus around the center of the frequency domain data set and zero all the terms that lie inside it. We can do this with a *Band Stop Filter* which is equivalent to the combination of an Ideal high pass filter with an Ideal low pass filter that has a slightly lower cutoff frequency (Fig. 7.11).

The advantage of the band stop filter over manual zeroing of the noise terms (as we did in Fig. 7.10) is that we do not have to know the exact orientation of the noise and search for the noise spikes in the Fourier spectrum. Because the range of frequencies removed is small, relative to the number of terms in the frequency domain, the ringing artifact associated with the band stop filter will usually be small, despite the steep profile of the filter.

A band stop filter tool is not included in the standard ImageJ package. It can be implemented by manual editing of frequency domain data followed by inverse Fourier transformation. Menu: Process > FFT > FFT will Fourier transform

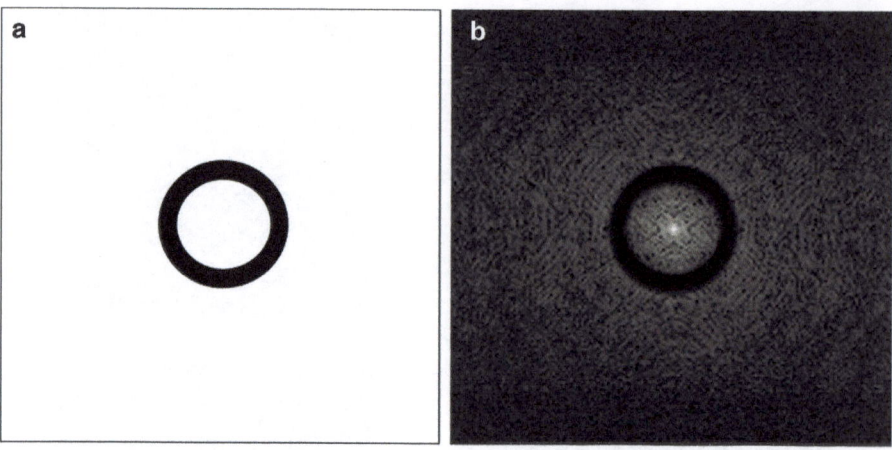

**Fig. 7.11** An Ideal *Band Stop* filter mask (**a**) applied to the frequency domain data of Fig. 7.10 (**b**)

an image and display the Fourier spectrum. The frequency domain data can be edited by applying the region selection tools on the Fourier spectrum. Menu: Process > FFT > Inverse FFT performs an inverse Fourier transform of the edited frequency domain data and displays the magnitude image.

### 7.2.5 Band Pass Filters

The complement of the band stop filter is the *Band Pass Filter*. It is effectively an annular mask, usually wider than a band stop mask, which attenuates all amplitudes outside the annulus (Fig. 7.12). Band pass filters are typically used to select 'soft' edge detail (intermediate frequencies) while simultaneously reducing high frequency noise.

The simplest, 'Ideal', band pass filter completely suppresses spatial frequencies above the upper and below the lower cutoff frequencies and has no effect on spatial frequencies inside the *pass band* (Fig. 7.12a). The sharp cutoffs mean this filter will tend to produce the same ringing artifact problems as an Ideal high or low pass filter. This is not surprising since this band pass filter mask is simply the elementwise product of Ideal low pass and high pass masks where the cutoff of the low pass mask is at a higher spatial frequency than the cutoff of the high pass mask.

A more versatile implementation of the band pass filter would use a mask for which we can control both the cutoff frequencies *and* the slope of the profile. A gradual gradient in the profile, as illustrated in Fig. 7.12b, will reduce or prevent ringing artifacts. In Fig. 7.13 a small amount of ringing artifact is present. To completely

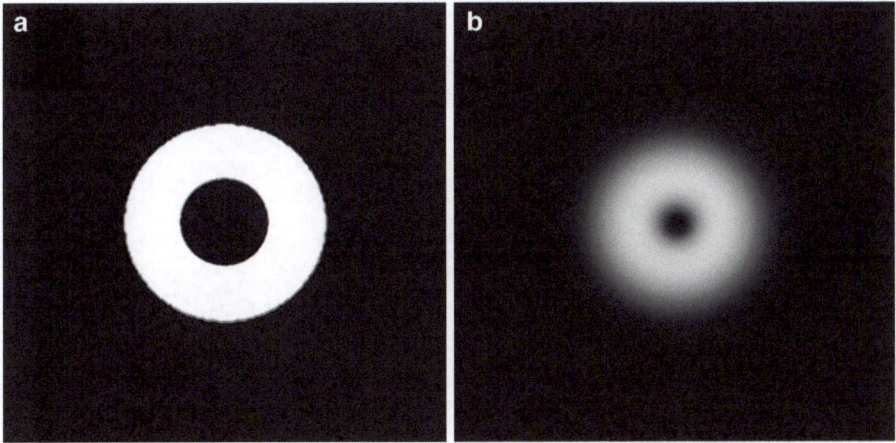

**Fig. 7.12** Band pass filter masks. Mask **a** has sharp ('Ideal') cutoffs and is likely to produce severe ringing artifacts when applied to an image that contains edge details. Filter **b** has much more gradual cutoffs and is less likely to produce a ringing artifact. The effects of these two filters are illustrated in Fig. 7.13 below

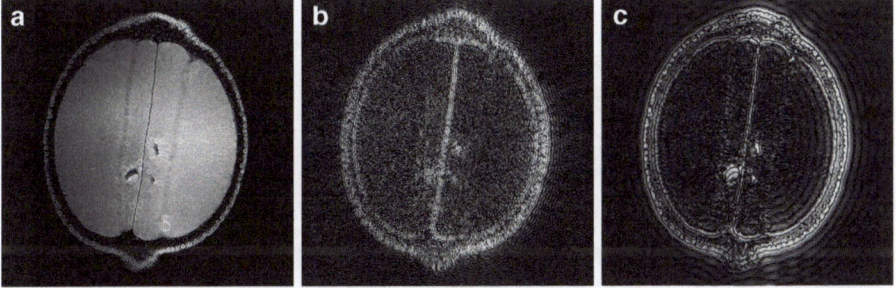

**Fig. 7.13** Illustration of the effects of the band pass filter masks shown in Fig. 7.12 above. (**a**) Original image. (**b**) Effect of Ideal band pass filter. (**c**) Effect of more gradual filter. Note that the pass band for the Ideal filter is above that for the more gradual filter, with the result that the gradual filter has retained more tonal information from the original image. Some ringing artifact is present in image **c** and it has a lower frequency (greater distance between lines) than the artifact in image **b**

eliminate the ringing artifact we need to use a filter with a much more gentle cutoff, as demonstrated in Fig. 7.14.

Unless the pass band of a band pass filter is very wide there may be a very significant loss or suppression of spatial frequency information from the original image. We have seen how attenuation of low frequency information with a high pass filter removes tonal information from an image but retains edge information. Conversely, a low pass filter will remove only high frequency information while low frequency tonal information is retained. The lost edge information results in a blurring of the image. A band pass filter with a narrow pass band can produce a

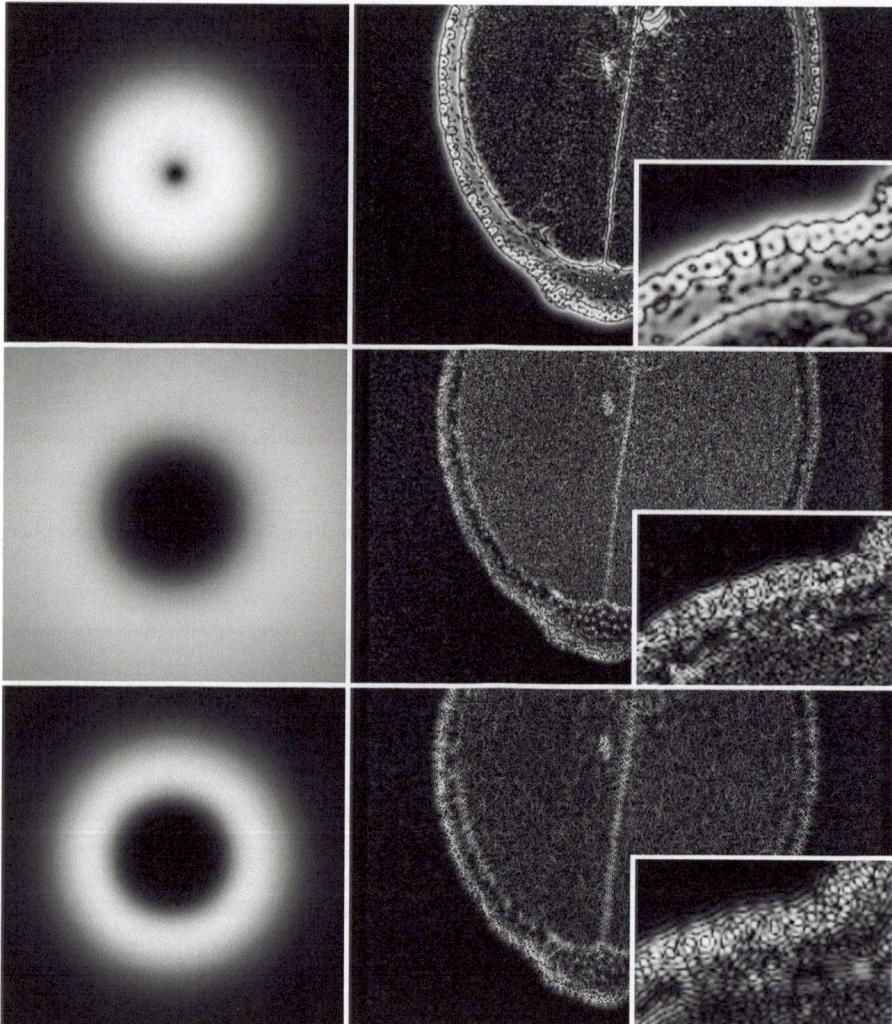

**Fig. 7.14** Illustration of the effects of three different band pass filter masks. The inset images show enlarged detail. Only the *top mask* retains any tonal detail from the original image. In contrast the middle and lower masks suppress all medium and low frequency information with the result that only edge detail is retained. The *bottom mask* has a much lower high frequency cutoff than the *middle mask* and this results in a high frequency ringing artifact

filtered image with major distortions due to the large amount of spatial frequency information that has been suppressed. This outcome is illustrated in Fig. 7.15 where we see that a narrow pass band results in distortions of the spatial domain data.

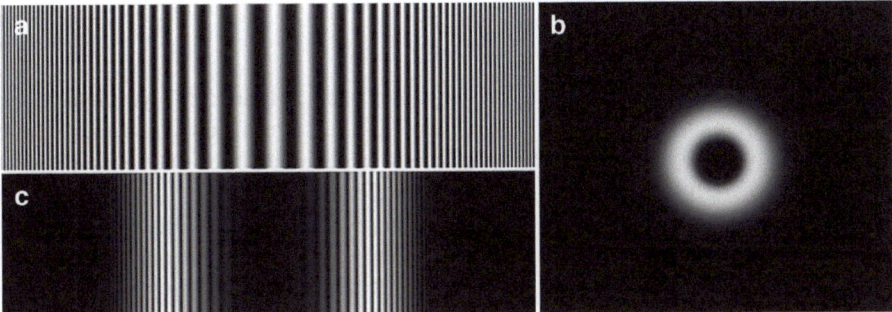

**Fig. 7.15** Illustration of some unexpected effects of a band pass filter. The synthetic image (**a**) contains spatial frequencies varying continuously from 16 cycles per image width at its center to 64 cycles at the left and right edges. Because the spatial frequencies vary continuously, rather than discretely, no region of the image can be described by a single spatial frequency. Application of the band pass filter (**b**) results in suppression of the highest and lowest spatial frequencies as expected (**c**), however, the retained intensity pattern is distorted because part of the original frequency information has been attenuated

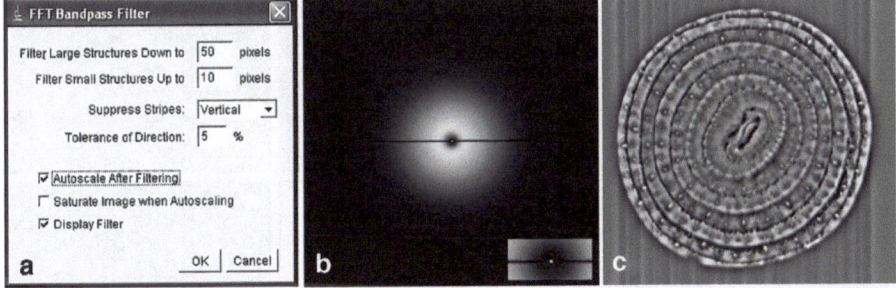

**Fig. 7.16** Application of the ImageJ band pass filter to the periodic noise-affected image from Fig. 7.10. The filter parameters (**a**) are described in terms of the size of the image structures that are attenuated, rather than spatial frequencies. This filter also enables the suppression of vertical or horizontal stripes by addition of a directional band stop filter. In this example the vertical stripes filter is implemented and appears as the horizontal black line in the center of the filter mask (**b**). The inset shows an enlargement of the center of the filter mask. Note that in this mask the central element is white so the zero frequency amplitude remains unchanged by the filter – the result is that the average signal intensity in the image is preserved. Note also that the periodic noise is incompletely suppressed (**c**) because the narrow horizontal black line in the filter mask is narrower than the white periodic noise spots (Fig. 7.10b)

The ImageJ tool Menu: Process > FFT > Bandpass Filter... implements an adjustable band pass filter with Gaussian profiles (Fig. 7.16). This filter also enables the suppression of vertical or horizontal stripes by addition of a directional band stop filter (see Fig. 7.16).

## 7.2.6 Directional Filters

All the filters we have looked at so far have been rotationally symmetrical or *isotropic* – their effect on spatial frequencies is the same in all directions. There is no requirement for filters to be isotropic and occasionally we may wish them to be otherwise. We might, for example, want to select only the vertical edges from an image or perhaps blur all the vertical edges. Figures 7.17 and 7.18 illustrates how we could do either of these things with an *anisotropic* directional filter. In this example the effect of directional low pass and high pass filters is shown. The profile of these two filters is Gaussian so the filtering occurs without ringing artifact.

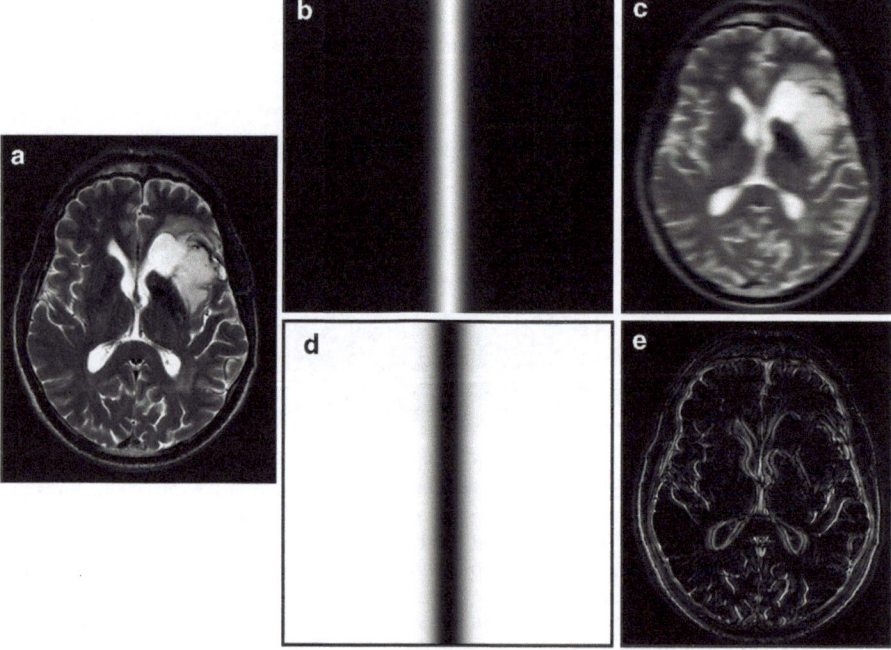

**Fig. 7.17** The effect of directional (anisotropic) filter masks. The x-low pass filter (**b**) suppresses high and intermediate frequencies in the *x*-direction: The vertical edges in the filtered image (**c**) are blurred, but horizontal edges are unaffected. The x-high pass filter (**d**) suppresses *all* low frequencies, and *y*-direction high and intermediate frequencies – those encoding horizontal edges: All tonal details, and all horizontal edges in the filtered image (**e**) are suppressed. Note that, because filter **b** is the complement of filter **d**, addition of the two filtered images (**c**, **d**) would produce the original unfiltered image (**a**)

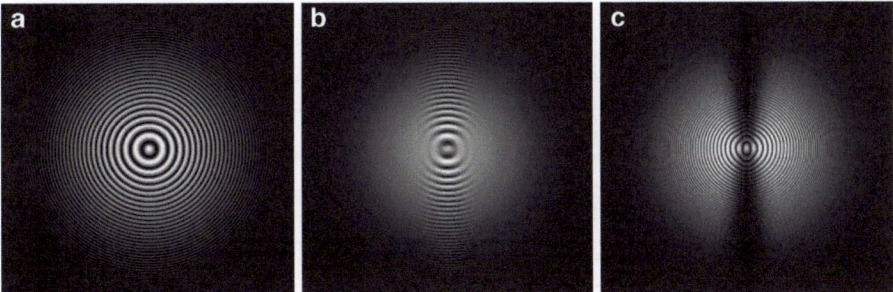

**Fig. 7.18** The effect of the directional high and low pass filter masks of Fig. 7.17 on a synthetic image. The test image (**a**) has spatial frequencies ranging continuously from 16 cycles per image width at the center, to 128 cycles at the edges. The effect (**b**) of the x-low pass filter (Fig. 7.17b) is to blur all edges that are *not* horizontal. Only the lowest frequency vertical components, at the center of the image, are not significantly blurred. In contrast, the effect (**c**) of x-high pass filter (Fig. 7.17d) is to suppress all horizontal edges. As above, addition of the two filtered images would produce the original unfiltered image

## 7.3 Spatial Domain Filters

Instead of transforming an image to the spatial frequency domain and performing a global filtering operation as is done with the frequency domain filters, we can process an image directly in the spatial domain according to particular characteristics of its pixel intensities. In this case the pixels of the image are processed individually, with changes made according to the properties of a pixel and its neighbors. Spatial domain filtration is thus often described as a *neighborhood operation*.

### 7.3.1 Smoothing and Blurring

Consider a noisy image. If the noise is, or can be expected to be, Gaussian, then we know that the intensities of all pixels differ positively or negatively from their expected noise-free values by an amount that has a Gaussian distribution and thus a zero mean. We can reasonably argue that if we adjust a pixel's intensity according to the mean intensity in its local neighborhood, then the adjusted intensity will be closer to the expected noise-free intensity for that pixel. The simplest way to do this adjustment is to change every pixel's intensity to the average intensity of that pixel and its near neighbors. This is called a *smoothing* operation.

In the simplest implementation (Fig. 7.19) of a smoothing operation each pixel's original intensity is replaced with the *average* of the nine pixels in the 3 × 3 neighborhood. This averaging process produces a *new image* with reduced variance in the local intensities, in other words, it is a smoothed version of the original image. It is important to note that the new pixel intensities are written into a new image matrix,

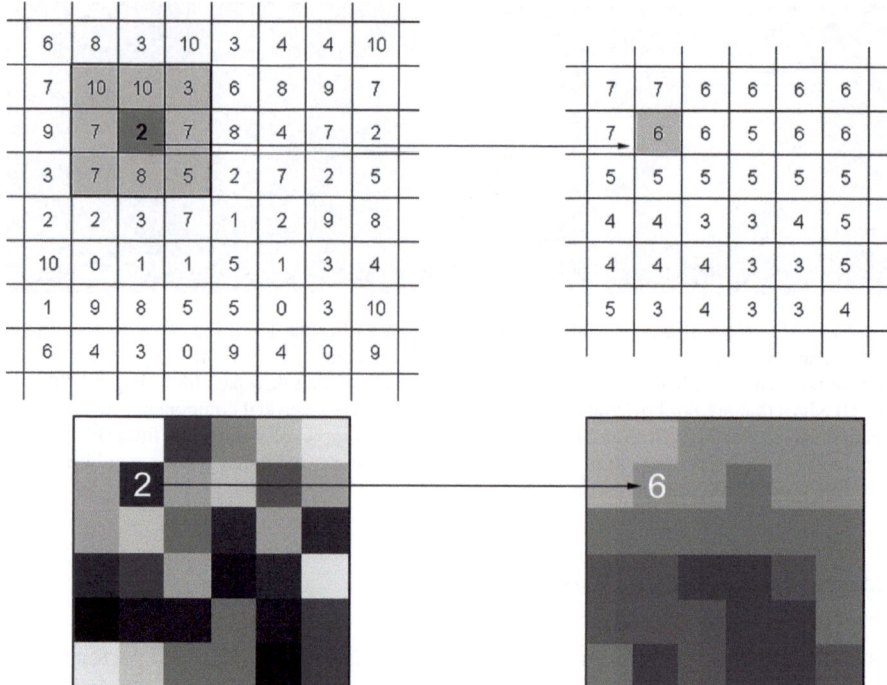

**Fig. 7.19** Demonstration of smoothing by averaging in a 3 × 3 neighborhood. The *top left* diagram represents part of the pixel intensity matrix of the image to be smoothed. The *shaded area* represents the local 3 × 3 neighborhood of the pixel who's original intensity is 2. The new intensity value for this pixel is calculated by taking the average intensity of all nine pixels in the shaded area. This new intensity value is written into the *new* image matrix shown in the top right diagram. The *bottom* images show a gray scale representation of the intensity data above. (We have shown only the inner region values in the new image matrix. The problem of what do when the neighborhood extends beyond the edge of the original image data is described in the text.) In these two images we have scaled the intensities so that $0 = black$ and $10 = white$. Note how the smoothing operation reduces the local contrast

not the original matrix. This prevents the adjusted value of one pixel being used in calculation of the new values for its neighbors.

The local averaging process just described is a specific example of the general local neighborhood process termed *Convolution*. Our example could be described as the repeated application of a 3 × 3 mask, conventionally called a *kernel*, in which each element of the kernel is $\frac{1}{9}$ as shown in Fig. 7.20. It is conventional to describe the elements of kernels as positive or negative integers rather than fractions, and to divide the output by the sum of all the elements (note that the output is not scaled when the sum of the elements is zero, as is the case for the Laplacian kernel described later).

Obviously, we are not limited to 3 × 3 kernels filled with identical elements. A greater smoothing effect would be achieved by using a 5 × 5 kernel, in which each element was 1 (scaled by 25). This would average the intensities of 25 pixels in the neighborhood and, while producing a smoother image, would also have the (usually

| | | | | | | | |
|---|---|---|---|---|---|---|---|
| 1/9 | 1/9 | 1/9 | | 1 | 1 | 1 |
| 1/9 | 1/9 | 1/9 | = 1/9 × | 1 | 1 | 1 |
| 1/9 | 1/9 | 1/9 | | 1 | 1 | 1 |

**Fig. 7.20** A 3 × 3 averaging convolution kernel as used in the process illustrated in Fig. 7.19. It is conventional to describe the elements of kernels as positive or negative integers, rather than fractions, and to divide the output by the sum of all the elements (except when the sum of the kernel elements is zero, in which case we don't scale the output)

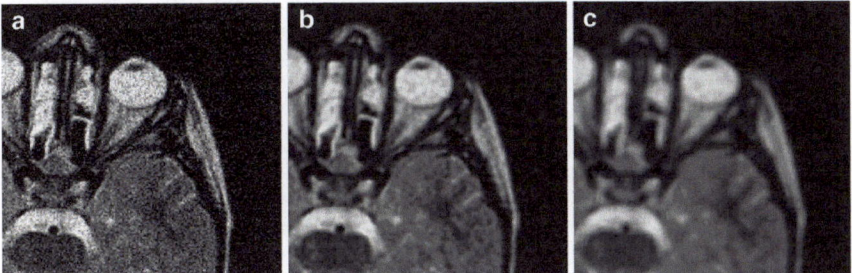

**Fig. 7.21** Illustration of the effect of simple averaging on image noise. (**a**) Noisy image. (**b**) Effect of 3 × 3 averaging kernel. (**c**) Effect of 5 × 5 averaging kernel. Although the noise is reduced there is significant blurring

undesirable) effect of reducing the sharpness of any edge detail in the image. A comparison of the effects of 3 × 3 and 5 × 5 averaging kernels is illustrated in Fig. 7.21.

The application of a convolution kernel differs distinctly and importantly from the application of a frequency domain filter mask. The convolution process can be described in stepwise fashion as follows:

1. Center the kernel over the first pixel in the image to be filtered.
2. Multiply each kernel element value by the pixel intensity that currently lies beneath it.
3. Add all the products from the previous step together.
4. Scale the output: Divide the sum of the products by sum of the kernel elements (If the sum of the kernel elements is zero then do not scale the output).
5. Write the scaled value into the new image matrix in the position corresponding to the pixel currently lying under the center of the kernel.
6. Move the kernel so it is centered over the next pixel in the original image.
7. Repeat steps 2–6 above for every pixel in the image.

This process is illustrated in Fig. 7.22. (Strictly speaking this process is *correlation*, not convolution. The difference, which mostly doesn't matter, is described at the end of this chapter.)

| A | B | C |
|---|---|---|
| D | E | F |
| G | H | I |

KERNEL

$$m' = \frac{Ag + Bh + Ci + Dl + Em + Fn + Gq + Hr + Is}{A+B+C+D+E+F+G+H+I}$$

| Original Image matrix | | | | | | | New Image matrix | | | | |
|---|---|---|---|---|---|---|---|---|---|---|---|
| a | b | c | d | e | | | a' | b' | c' | d' | e' |
| f | $A_g$ | $B_h$ | $C_i$ | j | | | f' | g' | h' | I' | j' |
| k | $D_l$ | $E_m$ | $F_n$ | o | | | k' | l' | m' | n' | o' |
| p | $G_q$ | $H_r$ | $I_s$ | t | | | p' | q' | r' | s' | t' |
| u | v | w | x | y | | | u' | v' | w' | x' | y' |

**Fig. 7.22** A diagrammatic representation of the process of convolution with a 3 × 3 kernel

**a**       **b**       **c**

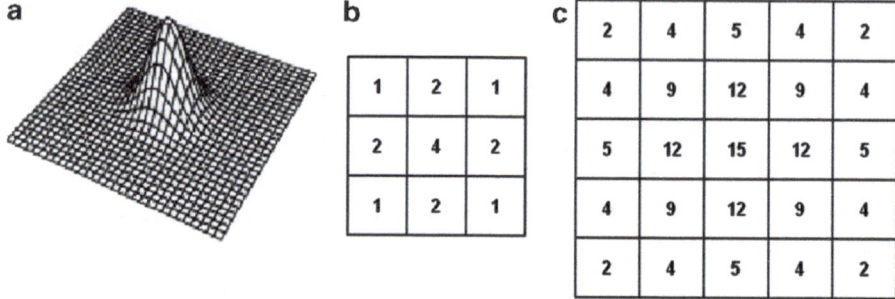

| | | | | |
|---|---|---|---|---|
| 2 | 4 | 5 | 4 | 2 |
| 4 | 9 | 12 | 9 | 4 |
| 5 | 12 | 15 | 12 | 5 |
| 4 | 9 | 12 | 9 | 4 |
| 2 | 4 | 5 | 4 | 2 |

| | | |
|---|---|---|
| 1 | 2 | 1 |
| 2 | 4 | 2 |
| 1 | 2 | 1 |

**Fig. 7.23** Gaussian convolution kernels are *discrete* approximations to a 2D Gaussian profile. (**a**) Perspective view of a 2D Gaussian profile. (**b**) 3 × 3 approximation to **a**. (**c**) 5 × 5 approximation to **a**. The effects of convolution with these two kernels are illustrated in Fig. 7.24

    The severity of the smoothing effect of a simple averaging kernel can be reduced by applying more weight to the central kernel element(s). This will mean the new value for a pixel will be less strongly affected by the value of its neighbors. For larger kernels the weighting usually decreases with distance from the center. A common weighting function corresponds to an approximation to a 2D Gaussian or bell curve (Fig. 7.23). This kernel is a *discrete approximation* to the continuous 2D Gaussian function that we saw previously as the Gaussian frequency domain filter (Fig. 7.23).

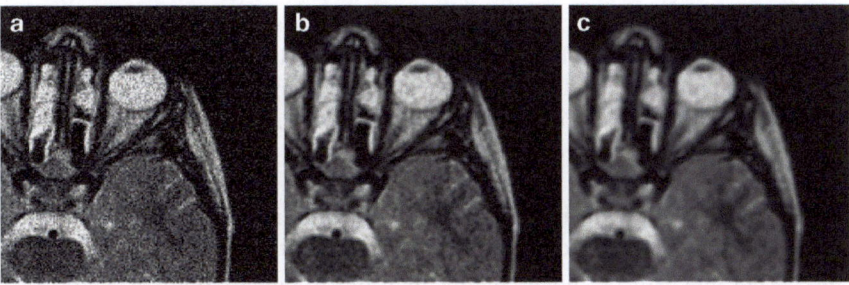

**Fig. 7.24** Use of Gaussian kernels from Fig. 7.23 to reduce image noise. (**a**) Noisy image. (**b**) Effect of $3 \times 3$ Gaussian kernel. (**c**) Effect of $5 \times 5$ Gaussian kernel. Compared with Fig. 7.21 the Gaussian kernels are less effective in reducing noise than the simple averaging kernels, however, they produce much less image blur for the same kernel size

Smoothing using a convolution kernel of this type is termed Gaussian blur (Fig. 7.24). The amount of blur is controlled by adjustment of the $\sigma$ parameter which represents the radius in pixels of the Gaussian profile at 0.6 of its maximum height.

A reasonably accurate approximation to a Gaussian profile of large radius will require a kernel of size at least four times the $\sigma$ of the Gaussian. As the size of a kernel increases the number of calculations required increases drastically. This can make the application of large kernels very slow. For a large convolution kernel it is often more computationally efficient to convert the kernel into its frequency domain equivalent and perform the filtering in the frequency domain.

The smoothing and blurring kernels result in a loss of information from the image because initially separate intensity information is spread out into the neighborhood. This is good for noise reduction but bad for spatial resolution. The endpoint of severe or repetitive blurring would be an image of completely uniform intensity containing no useful spatial or contrast information – the only information left would be the average intensity of the original image.

A practical problem arises when we attempt to apply a convolution kernel to the edge pixels of a spatial domain image matrix. We have to make a decision about what to do when parts of the kernel 'overhang' the edges of the image. One solution is to ignore the kernel elements that are outside the image and adjust the scaling of the output according to the sum of the elements that do overlie the image. An alternative, slightly bizarre, solution is to use the gray scale intensities from the opposite side of the image. This should sound familiar. It is the same as treating the image as if it were tiled in a 2D plane – just as the Fourier transform treats the spatial domain data as if it were tiled infinitely. In the case of convolution we do not need infinite tiling – nine tiles is sufficient to perform all of the necessary calculations on the central tile.

The smoothing or blurring filters are called low pass filters because they attenuate high spatial frequencies and they pass low frequencies without attenuation. We could see this effect if we compared the Fourier spectra of an image before and after application of a smoothing kernel. Alternatively, we could *deduce* this effect by performing a Fourier transform of the kernel itself and looking at its Fourier spectrum. It turns out that, mathematically, the application of a convolution kernel in the spatial domain is equivalent to the multiplication of the FT of the image by the FT of the convolution kernel, followed by inverse Fourier transformation to get back to the spatial domain. We need not concern ourselves with the maths here but this is a very useful property. For large convolution kernels processing can be faster if we convert the kernel into its frequency domain equivalent and do the processing in the frequency domain.

An initial problem with simply performing a Fourier transform on a convolution kernel is that the resultant frequency domain matrix will have the same dimensions as the kernel and will be much smaller than the frequency domain matrix of the image to be filtered. Remember that the frequency domain filter mask must have the same dimensions as the image matrix. The way to get around this problem is to first increase the size of the kernel by surrounding (padding) it with zeros until it has the same pixel dimensions as the image to be filtered. This will not alter the effect of the convolution filter, even if the enlarged zero-padded version were used in the spatial domain convolution process. Figure 7.25 illustrates the Fourier spectra of the simple 3 × 3 averaging kernel (Fig. 7.20) and the Gaussian kernels shown in Fig. 7.23. As expected the Fourier spectra demonstrate that the averaging and Gaussian kernels act as low pass filters and the profiles of the Fourier spectra are consistent with the different degrees of blurring expected from these kernels. In the Fourier spectrum of the simple averaging kernel we see the suggestion of a ringing artifact. This feature is expected in the Fourier spectrum as a result of the step-shaped profile of the kernel, and it suggests that the highest spatial frequencies will not be 100% attenuated. That this is indeed the case is illustrated in Fig. 7.26.

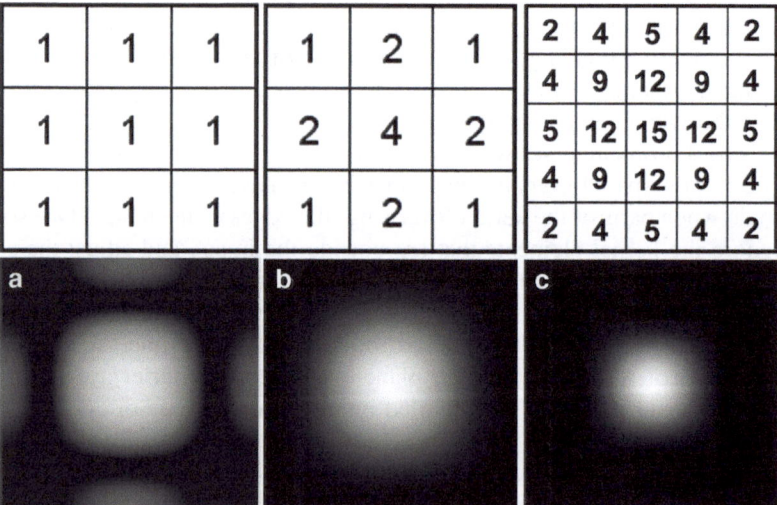

**Fig. 7.25** Averaging and smoothing kernels and their Fourier spectra. The Fourier spectrum of the simple averaging kernel (**a**) shows significant amplitudes at high spatial frequency. This is a result of the step-shaped profile of the kernel and it suggests that high spatial frequencies will not be 100% attenuated. This is true. The effect is an *aliasing artifact* and is demonstrated in Fig. 7.26

**Fig. 7.26** Demonstration of the production of *aliasing* artifact by the simple 3×3 averaging kernel. The test image **a** comprises one pixel wide white lines spaces one pixel apart. Convolution with the 3 × 3 averaging kernel (Fig. 7.25a) does not smooth the image but instead produces a set of aliased lines (**b**). This effect can be predicted from the non-zero amplitudes of high spatial frequencies in the Fourier spectrum of the kernel (Fig. 7.25a). In contrast, convolution with the 3 × 3 Gaussian kernel (Fig. 7.25b) completely smoothes the line pattern (the dark extremities of the filtered images result from zero padding of the image edges prior to convolution)

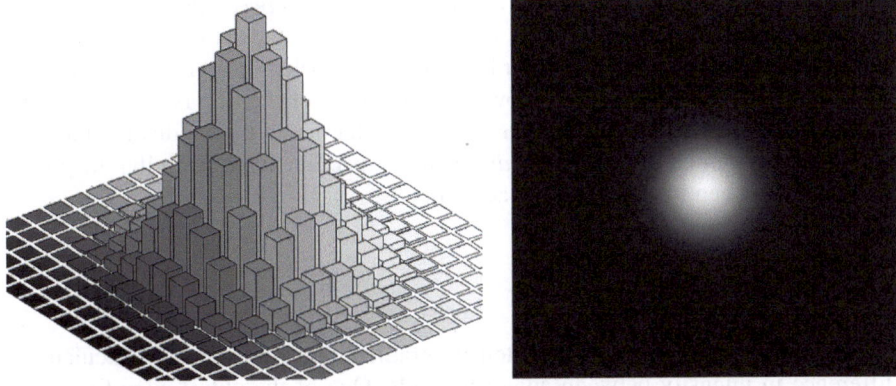

**Fig. 7.27** A 3D plot of a 17 × 17 Gaussian smoothing kernel and its Fourier spectrum. In this example the size of the kernel is sufficient to ensure that its Fourier spectrum is fully symmetrical

We mentioned above that a special property of the Gaussian shape is that its Fourier spectrum is another Gaussian. Figure 7.27 shows a 17×17 Gaussian smoothing kernel and its Fourier spectrum. The Fourier spectra of the smaller Gaussian kernels illustrated in Fig. 7.25 are somewhat distorted from the expected circular shape due to the relatively poor approximation of the small kernels to the true Gaussian profile.

Use the ImageJ tool Menu: Process > Filters > Convolve... to create and test averaging and Gaussian kernels (or any arbitrary kernel). Alternatively, the tool Menu: Process > Filters > Mean... creates and applies *roughly* circular averaging kernels with a user-specified radius ($r = 1$ gives a 3 × 3 kernel, $r = 2$ gives a 5 × 5 kernel with corner elements set to zero, and so on).

## 7.3.2  Gradients and Edges

Human perception has a strong focus on edge detail and we tend to regard images containing well-defined sharp edges as being of superior quality to blurred images. Edge enhancement of an image assists us in extracting visual information from the image. Another common application of convolution filters is the extraction or enhancement of edge detail.

We have already seen how a high pass filter in the frequency domain can be used to select the edge details in an image. By making the filter asymmetrical, or anisotropic, edges that have a particular orientation can be selected. The same sorts of edge selection can also be performed in the spatial domain with convolution filters, however, since we are now using a neighborhood operation we have to think about the *local* properties that characterize an edge.

A simple definition of an edge in an image is a *'line between adjacent regions of distinctly different intensity'*. Notice that this definition does not say how wide the line is, nor what its orientation is, nor how big the intensity difference is between the adjacent regions. Intuitively we know that the narrower the line is, and the bigger the difference between the adjacent regions, the sharper and more distinct the edge is. The most significant term here is *difference* – specifically, for a digital image, the difference between intensities of adjacent pixels.

### 7.3.2.1  Roberts Cross Filter

The simplest way to measure the intensity gradients in an image is to calculate the difference in intensity between adjacent pixels. One of the oldest edge filters, the Roberts Cross filter, does this with the pair of $2 \times 2$ convolution kernels shown in Fig. 7.28. Convolution with each of these kernels effectively calculates the *first derivative*, in other words the *gradient*, of the intensity along diagonals. The output from convolution with either of these kernels will be positive or negative depending on the direction of the intensity gradient, as illustrated in Fig. 7.29. In looking for edges in an image we are normally more interested in the magnitude of the gradient ($G$) than its sign or direction so we *add* the absolute values of the outputs $g_{R1}$ and $g_{R2}$ from each convolution:

$$G = |g_{R1}| + |g_{R2}| \tag{7.3}$$

The value of $G$ at each pixel will then have a possible range from zero (all diagonally adjacent pixel intensities equal) to a maximum equal to twice to the maximum pixel intensity.

The Fourier spectra of the two Roberts Cross kernels confirm that they act as directional high pass filters. This is just what we should expect since tonal (low spatial frequency) details of an image are regions of similar or identical intensity. The difference in intensity between adjacent pixels is small, so these regions give only a small or zero output when convolved with either of the Roberts Cross kernels. Conversely, edges or steep intensity gradients in the image will give a large output.

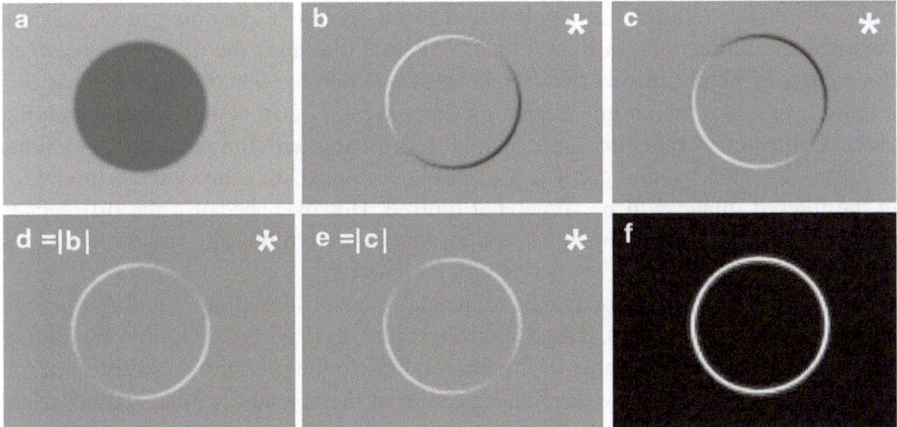

**Fig. 7.28** A very simple gradient filter simply measures the intensity difference between adjacent pixels. However, every pixel can have up to eight neighbors. The *Roberts Cross* filter kernels, shown here with their Fourier spectra, measure the intensity difference in two of the eight possible directions. The magnitude of the two outputs is normally added to give an *estimate* of the local gradient: $G = |g_{R1}| + |g_{R2}|$. The Fourier spectra indicate that the kernels act as directional high pass filters

**Fig. 7.29** Application of the Roberts Cross filter kernels (Fig. 7.28) to a test image (**a**). The individual kernels have high directional sensitivity as illustrated by their different outputs $(g_{R1})$ (**b**) and $(g_{R2})$ (**c**). It is conventional to convert the individual convolution outputs to absolute values (**d**, **e**) and to sum these to form the final output $(G = |g_{R1}| + |g_{R2}|)$ (**f**), (* Note that images **b**–**e** represent data containing negative values. In these four images mid-gray = 0, negative values are dark, and positive values light)

One of the drawbacks of the Roberts Cross filter is the fact that, because the $2 \times 2$ kernels have no single central element, it is not obvious where we should put the calculated output of the filter in the new image. Whatever choice is made, the filtered image features will be effectively shifted by $\frac{1}{2}$ pixel in both horizontal and vertical directions relative to their position in the original image. This filter also tends to exaggerate noise because all differences between diagonally adjacent pixels that are due purely to noise will contribute to the filter output.

Use the ImageJ tool Menu: Process > Filters > Convolve... to create and test the individual kernels of the Roberts Cross filters. Images must first be converted to 32-bit mode (Menu: Image > Type > 32-bit) as this is the only format in ImageJ that supports negative intensity values. For image display the range of values is scaled so that the minimum is displayed as black and the maximum as white.

### 7.3.2.2  Prewitt and Sobel Filters

The noise sensitivity of the Roberts filter can be reduced by measuring the average intensity gradient over a small local region, rather than between immediately adjacent pixels. A simple implementation of this gradient averaging process is the Prewitt kernel illustrated in Fig. 7.30. This kernel measures the difference between *three pairs* of pixels spaced two pixels apart and outputs the *sum* of the three differences. Just like the simple averaging kernel (Fig. 7.20), or any other averaging process, this kernel blurs the input spatial information with the result that gradients or edges detected by this method are poorly localized. In other words, the edges in the filtered image will be somewhat blurred. However, because the differences are measured on a larger local scale than the Roberts filter, any large differences between immediately adjacent pixels will be ignored.

Blurring in edge detection can be reduced, just as in smoothing, by *weighting* the gradient averaging process. The Sobel filter kernels (Fig. 7.31) do this by adding weight to the central elements of the kernel. Due to the direction-sensitivity of the kernel, four orientations are used and the sum of the absolute values is used as the output:

$$G = |g_x| + |g_{xy}| + |g_y| + |g_{yx}| \tag{7.4}$$

or, alternatively:

$$G = \sqrt{g_x{}^2 + g_{xy}{}^2 + g_y{}^2 + g_{yx}{}^2} \tag{7.5}$$

**Fig. 7.30** The basic Prewitt filter kernel. In contrast to the Roberts Cross filter this kernel calculates the *average* gradient in a small region of the image rather than the gradient between immediately adjacent pixels

| -1 | 0 | 1 |
|----|---|---|
| -1 | 0 | 1 |
| -1 | 0 | 1 |

| 1 | 2 | 1 |
|---|---|---|
| 0 | 0 | 0 |
| -1 | -2 | -1 |

$g_y$

| 2 | 1 | 0 |
|---|---|---|
| 1 | 0 | -1 |
| 0 | -1 | -2 |

$g_{yx}$

| 1 | 0 | -1 |
|---|---|---|
| 2 | 0 | -2 |
| 1 | 0 | -1 |

$g_x$

| 0 | -1 | -2 |
|---|---|---|
| 1 | 0 | -1 |
| 2 | 1 | 0 |

$g_{xy}$

**Fig. 7.31** The basic Sobel filter kernels are similar to the Prewitt kernels but calculate a *weighted average* of the gradient in a small region of the image, and thus output edges with less blurring than the Prewitt kernel

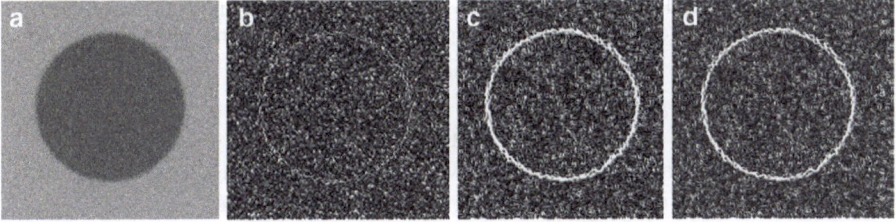

**Fig. 7.32** Comparison of the edge and noise sensitivities of the Roberts (**b**), Prewitt (**c**), and Sobel (**d**) gradient detection filters applied to a test image (**a**). The edge sensitivity of the Roberts Cross filter is poor compared with its noise sensitivity. The Sobel filter produces a slightly better defined edge than the Prewitt filter

A variant of the Sobel filter is the Kirsch filter which applies the four Sobel filter kernels and selects the largest of the four gradient magnitudes as the output for each pixel.

The relative sensitivities to gradients and image noise of the Roberts, Prewitt, and Sobel filters is illustrated in Fig. 7.32. The $2 \times 2$ kernel of the Roberts filter is unable to discriminate between a gradient and noise. The Prewitt and Sobel kernels are less sensitive to noise but produce a rather poorly localized description of the edge feature.

> The Sobel filter is implemented in the ImageJ Menu: Process > Find Edges command.

### 7.3.2.3 Laplacian Filters

The directional sensitivity of the above filters means it is necessary to apply rotated variants of the kernels several times and combine the outputs to get a direction-insensitive output. Often it is more convenient to use a more isotropic single kernel. The simplest and commonest of these is the Laplacian, the $3 \times 3$ kernel of which is shown in Fig. 7.33 together with its corresponding Fourier spectrum. The Laplacian

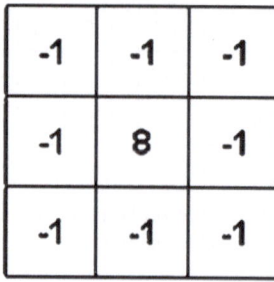
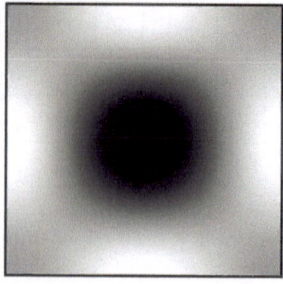

Fig. 7.33 A Laplacian kernel calculates the *second derivative* of the image intensity. The 3 × 3 Laplacian kernel and its Fourier spectrum are shown here. Note that the sum of the kernel elements is zero, so the output is not scaled. The Fourier spectrum confirms that convolution with this kernel acts as a high pass filter

is formally a *second derivative* filter – it measures, effectively, the gradient of the gradient. The advantage of a second derivative filter for edge detection is that it will usually define edges more precisely than a first derivative filter.

How can we tell that the Laplacian calculates the second derivative of pixel intensities? This is not immediately obvious looking at the kernel, but it makes sense if we think about what is happening in any single line through the center of the kernel: The sequence of kernel elements is −1,8,−1. In other words, the kernel is the sum of four 3 × 1 kernels (one vertical, one horizontal, and two diagonal) whose elements are [−1 2 −1]. Any one of these 3 × 1 kernels is the sum of two 2 × 1 difference kernels with elements [−1 1] and [1 −1], or the *difference* between two identical difference kernels with elements [−1 1] where the second kernel is displaced one pixel from the first. The 2 × 1 kernels calculate the intensity differences between one particular pixel and those on either side of it. Subtracting one offset kernel from the other gives the 3 × 1 kernel [−1 2 −1] that calculates the difference between the differences – in other words, the second derivative! In contrast to the Roberts Cross filter we now have a single central kernel element so there is no confusion about where to put the output.

The sum of the elements of the Laplacian is zero. If it is applied to a neighborhood in which all the gray scale intensities are identical the output will be zero. Thus any region of constant intensity will become black in the filtered image. When the intensity of the central pixel differs substantially from its neighbors, as may be the case for noise, the output of the Laplacian is very high because it sums *all* differences between a pixel and its eight neighbors. Although the Laplacian filter is good for edge detection it has a tendency to exaggerate lines and noise even more than edges. For this reason it is sometimes used as a point defect detector. This effect is demonstrated in Fig. 7.34 where we see a test pattern comprised of mid-gray features: a rectangle, a narrow line, and a series of very small (almost invisible) points comprised of single pixels. When a 3 × 3 Laplacian kernel is applied to this image the edges of the rectangle are isolated as expected. However, the line feature is enhanced more than the edges of the rectangle, and the previously very faint points become quite distinct.

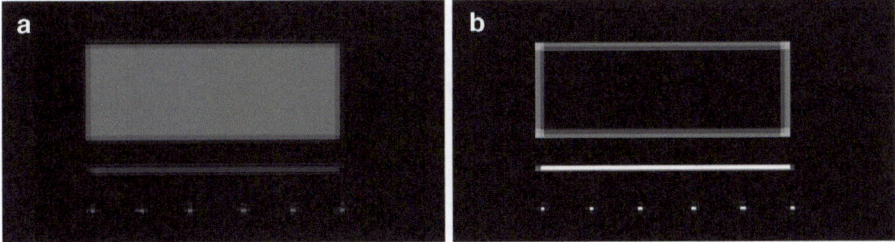

**Fig. 7.34** Demonstration of edge selection and point and line exaggeration of Laplacian kernel. (**a**) Test image. (**b**) Effect of convolution with 3 × 3 Laplacian. Note that the originally inconspicuous point features (single pixels of same intensity as large rectangle) are strongly enhanced, the line feature less so, and the edge of the gray rectangle less again. This is a notable characteristic of Laplacian filters

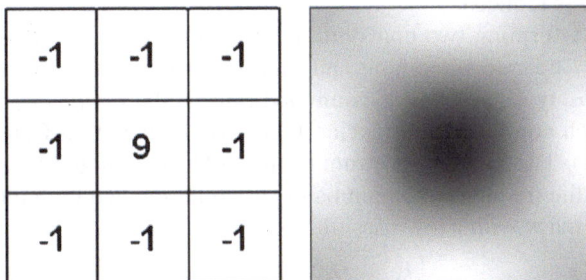

**Fig. 7.35** 3 × 3 'High Boost' kernel and its corresponding Fourier spectrum. Note that the sum of the kernel elements is one so the image intensity in homogeneous regions will be unchanged. The gray center of the Fourier spectrum indicates that low frequencies are only partially attenuated so tonal detail is retained

**Fig. 7.36** High Boost filter demonstration. (**a**) Original image. (**b**) Laplacian only. (**c**) High boost. Note that the high boost filter retains tonal information (the center of the rectangle)

Often the aim is not to isolate the edges in an image but to *enhance* them. This can be achieved by adding the output of a Laplacian high pass filter to the original image. This is equivalent to adding a one to the central element of the Laplacian convolution kernel (Fig. 7.35). The resulting filter is referred to as a *High Boost* filter because its effect is to increase the relative intensity of high spatial frequencies (Fig. 7.36).

Use the ImageJ tool Menu: Process > Filters > Convolve... to create and test the Laplacian kernels.

#### 7.3.2.4 LoG Filters

One way to get around the propensity of the Laplacian filter to exaggerate noise is to smooth the image before application of a high pass or high boost filter. In Fig. 7.37 we can see that smoothing prior to application of the high boost filter reduces the final intensity of the point features (representative of noise) and, to a lesser extent, the line feature. Note also that the degree of enhancement of the edge of the rectangle is reduced.

Instead of performing two convolution operations, the smoothing and then the high boost, the separate smoothing and high boost kernels can be combined into one kernel. Closely related to this combined filter is the *Laplacian of Gaussian* or LoG filter which combines a Laplacian high pass filter with a Gaussian low pass filter (Fig. 7.38). The Fourier spectrum of the LoG filter should look familiar. It is very similar to the frequency domain band pass filter created by multiplication of a high pass filter mask with a low pass filter mask.

**Fig. 7.37** Application of a smoothing filter prior to the high boost filter. (**a**) Original image. (**b**) Smoothed original. (**c**) High boost applied to image **b**. In comparison with Fig. 7.36 the point features, which simulate noise, are much less strongly enhanced

### 7.3.3 Spatial and Frequency Domain Properties of Convolution

The method of combining convolution filter kernels is convolution, *not* simple addition. Either of the two kernels can be treated as an image (to avoid any edge effects some zeros are added around the perimeter) and convolved with the other kernel. Multiplication of filter masks in the frequency domain is the same as convolution of kernels in the spatial domain. The relationship between the kernels and Fourier spectra shown in Fig. 7.38 can be described mathematically:

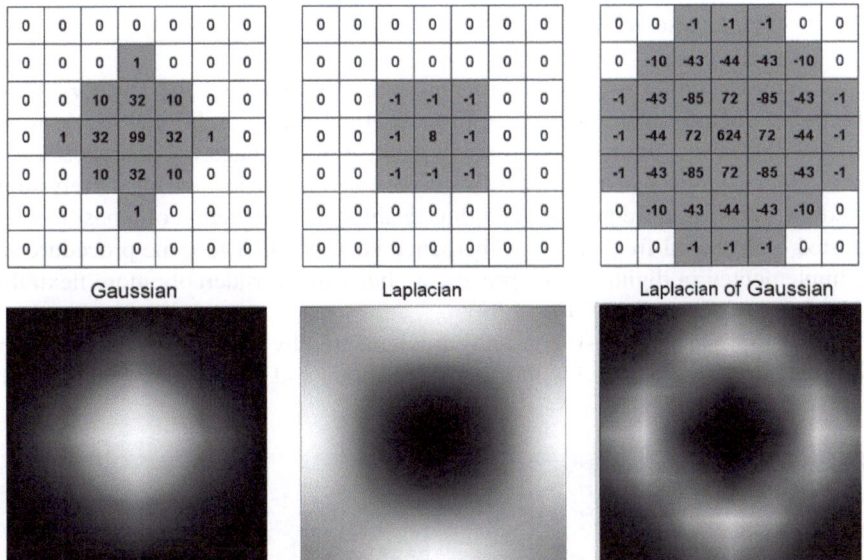

Fig. 7.38 Gaussian, Laplacian, and 'Laplacian of Gaussian' (LoG) kernels. The LoG kernel is created by convolution of the Gaussian and Laplacian kernels. The Fourier spectra are shown under the associated kernels. Note the similarity of the Fourier spectrum of the LoG kernel to the band pass filter mask of Fig. 7.12b, and that it can be formed by simple elementwise multiplication of the Fourier spectra of the Gaussian and Laplacian kernels. This equivalence is formalized in Eq. 7.6

$$K_{LoG} = K_G \otimes K_L = F'\big((F(K_G) \bullet F(K_L))\big) \qquad (7.6)$$

Where:

$K_L$ and $K_G$ are the Laplacian and Gaussian kernels
$K_{LoG}$ is the LoG kernel
$F$ denotes the Fourier transform
$F'$ denotes the inverse Fourier transform
'$\otimes$' is the convolution operation, and
'$\bullet$' is element-wise multiplication

This is simply a restatement of the mathematical relationship between convolution and Fourier transforms – the *Convolution Theorem*.

### 7.3.4 Convolution Versus Correlation

In strict mathematical terms the process we described as convolution (illustrated in Fig. 7.22) is actually *correlation*. Convolution differs from correlation in that the kernel matrix is rotated 180°. The difference is *only* significant for kernels that are not symmetrical when rotated 180°, and only then if we are interested in the raw output rather than the absolute or squared value. Most image processing texts use the term 'convolution' instead of 'correlation'.

### 7.3.4.1   The Unsharp Mask

A very common filter used for sharpening edges in images is the *Unsharp Mask*. The rather contrary name derives from its original use in traditional film printing where a deliberately blurred low contrast positive of an image was juxtaposed with the original negative to expose a new positive. The blurred positive effectively attenuated the low spatial frequencies and thus enabled the production of a positive with boosted high spatial frequencies – sharpened edge detail. The same procedure can be implemented in digital image processing but with considerably more flexibility and control of the outcome (Fig. 7.39).

The digital unsharp mask is usually implemented by a spatial domain convolution. The blurring is applied with a Gaussian kernel and the degree of blurring (the

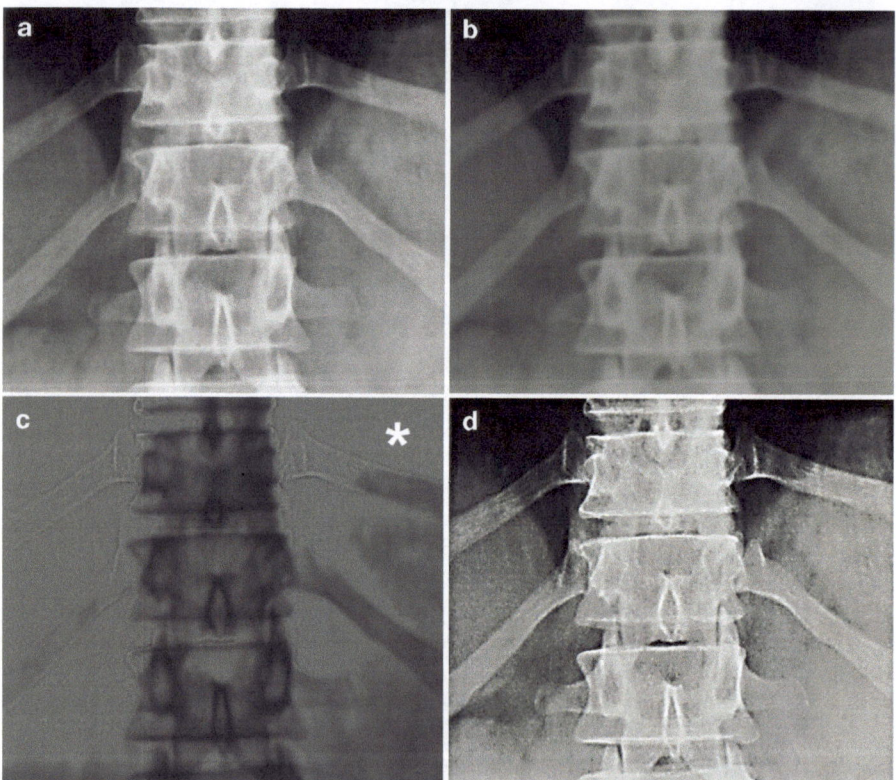

**Fig. 7.39** Illustration of the stepwise implementation of an unsharp mask (USM) for edge enhancement. (**a**) Image to be sharpened. (**b**) Gaussian blur applied to image **a**. (**c**) The difference image (**c**=**a**–**b**). (**d**) Sharpened image resulting from addition of images **a** and **c**. Image **b** is equivalent to the 'unsharp mask' originally used in film processing. In the digital implementation of the USM these separate steps are combined into a single mathematical operation (* Image **c** represents data containing negative values. In this image mid-gray $= 0$, negative values are dark, and positive values are light)

spatial extent of the sharpening adjustment) is controlled by adjusting the *radius* parameter (sigma in Eq. 7.2). The difference image is multiplied by a *weight* parameter to control the overall level of intensity adjustment. In the digital implementation of the USM all these steps and their parameters are combined into a single mathematical operation.

Since the unsharp mask is effectively a high boost filter it has a tendency to exaggerate noise and point features. Some implementations (e.g. Adobe Photoshop) also include an adjustable *Threshold* parameter to control this tendency. If the difference image value for a particular pixel does not exceed the threshold then no adjustment is made. Overuse of an unsharp mask filter will lead to conspicuous 'halo' artifacts along edges (Fig. 7.40).

We can demonstrate that the unsharp mask is equivalent to a high boost filter by rearrangement of Eq. 7.6. In this case we want to see the frequency domain filter mask that corresponds to a particular application of the unsharp mask. A little reorganization of Eq. 7.6 gives:

$$F(K_{USM}) = \frac{F(I \otimes K_{USM})}{F(I)} \tag{7.7}$$

where $I$ and $K_{USM}$ are the image and kernel matrices.

This expression tells us that the Fourier transform of the USM kernel ($F(K_{USM})$) is equal to the ratio of the Fourier transforms of the filtered image

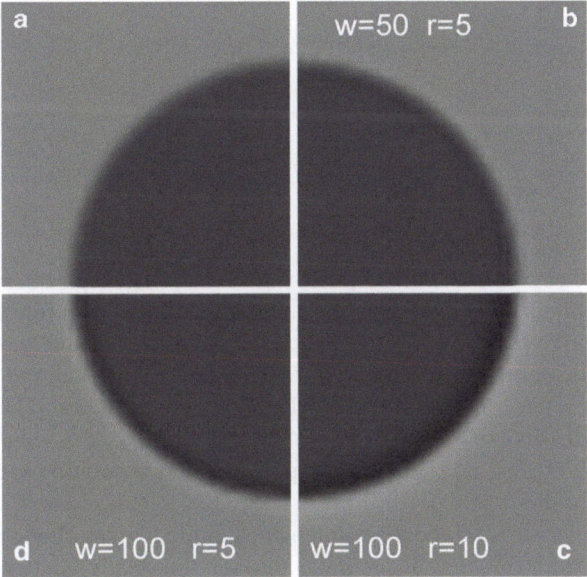

**Fig. 7.40** Artifacts due to overuse of the unsharp mask radius (r) and weight (w) parameters. Excessive weight produces a 'halo' along edges. The lateral extent of the halo increases with radius (sigma)

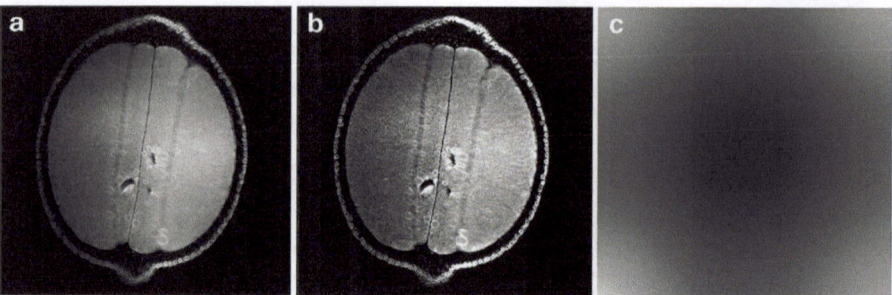

**Fig. 7.41** The unsharp mask is effectively a high boost filter. Although the filter kernel is not normally displayed its spatial frequency effect can be demonstrated. (**a**) Original image. (**b**) Image after application of USM with $w = 100$ and $r = 10$. (**c**) Frequency domain mask equivalent of the USM obtained by dividing the Fourier spectrum of image **b** by the Fourier spectrum of image **a** (Eq. 7.7)

and the original image. The magnitude of $F(K_{USM})$, is illustrated in Fig. 7.41c. As expected it has the appearance of a high boost mask. (This method is in reality not as simple as it looks because the Fourier spectrum of the original image contains many elements of amplitude zero and element-wise division by this matrix will cause problems. Adding a small positive value to all the amplitudes of $F(I)$ before the division operation gives a reasonable approximation to the desired output.)

> The ImageJ tool Menu: Process > Filters > Unsharp Mask... has controls for blurring (Sigma) and weight, but no threshold control.

### 7.3.5 Median Filters

When an image contains unwanted pixel values that differ significantly in intensity from their neighbors the application of a smoothing or blurring filter is only partially effective in removing these defects and the blurring effect is usually undesirable. Increasing the size of the averaging kernel, or the radius of the Gaussian blur kernel, in order to further reduce the magnitude of the pixel aberrations only further increases the blurring of the image. This is because the aberrant intensity values are included in the averaging process and skew the output. A common solution to this problem is to use a Median filter.

A median filter sorts all the pixel intensities in the neighborhood into numerical order and uses the value that lies *midway* through the ordered list as the new value for the central pixel position. The new value is then written into a new image matrix. No matter how different the intensity of the central pixel is from those of its neighbors it does not bias the intensity of the replacement.

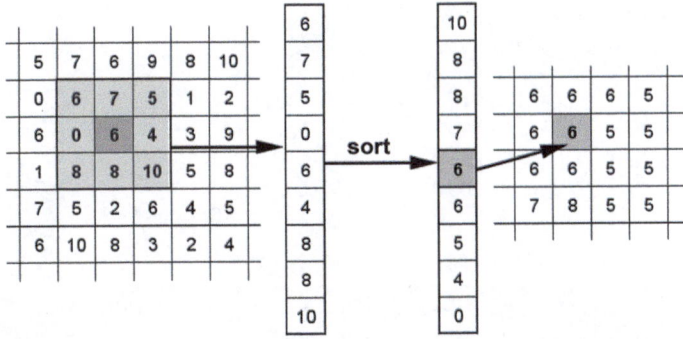

**Fig. 7.42** Diagrammatic representation of the median filter process. The intensities of all the image pixels that lie under the kernel are sorted. The median value is the one that lies at the center of the sorted list and this value is written into the new image matrix in the position that corresponds to the center pixel of the current kernel position. In this particular example the pixel intensity remains unchanged since its original value turned out to be the median of the neighborhood

The median filter process can be described in stepwise fashion as follows (Fig. 7.42):

1. Center the kernel over the first pixel in the image to be filtered.
2. Take all the pixel intensities that currently lie beneath the kernel and sort them in order of intensity.
3. Select the median intensity, which is the intensity that lies at the mid-point of the sorted list.
4. Write the median value into the new image matrix in the position corresponding to the pixel currently lying under the center of the kernel.
5. Move the kernel so it is centered over the next pixel in the original image.
6. Repeat steps 2–5 above for every pixel in the image.

The median filter is highly effective in reduction of point and line defects in images and has a smaller blurring effect than an averaging filter (Fig. 7.43).

The intensity sorting process involved in the Median filter make it very distinct from the averaging, blurring, and edge enhancement filters. There is no frequency domain equivalent to the sorting process because sorting is not a ('linear') convolution process. The median filter is a specific example of a *Rank Order* filter. All rank order filters are based on sorting of neighborhood pixel intensities. A Maximum filter uses the highest intensity as the new pixel value, while a minimum filter uses the lowest intensity. The rank order filters are examples of *non-convolution* or *non-linear* filters – they have no frequency domain equivalents.

## 7.3.6 Adaptive Filters

One problem with the median filter is that the only time it *does not* change a pixel's intensity is when the pixel happens to already have the median value in the neighborhood. This will still cause a small amount of blurring, though not as

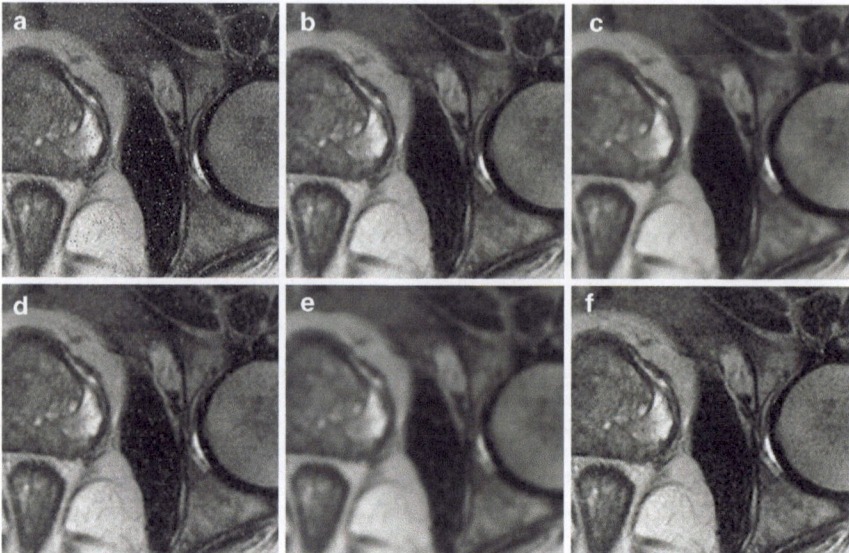

**Fig. 7.43** Comparison of median, smoothing, and adaptive filters for noise reduction. (**a**) Original image affected by both Gaussian and Salt&Pepper noise. (**b**) Median filter ($r = 1$). (**c**) Median filter ($r = 2$). (**d**) $3 \times 3$ averaging filter. (**e**) Gaussian blur ($r = 2$). (**f**) Adaptive filter ('Sigma Plus' filter plugin from ImageJ. $r = 1$. Only intensities greater than one SD from the local mean were changed)

severe as that caused by the averaging and Gaussian filters. What if the central pixel is not the maximum or minimum in the neighborhood or is not particularly different from the average of its neighbors? If we just want to eliminate intensities that are very different from those in the neighborhood then, instead of sorting the intensities, we could measure the amount of intensity variation in the neighborhood and *only* change a pixel's intensity if it exceeds some defined threshold of difference. An *Adaptive Filter* changes its behavior according to the current neighborhood properties.

A basic adaptive filter for removal of extreme pixel values might, for example, only change the value of a pixel if it is more than one standard deviation different from the mean of the neighborhood. Extreme values identified by this criterion could then be changed to the local mean value. The effect of such a filter is illustrated in Fig. 7.43f. Notice that it is extremely effective in removal of the salt and pepper noise but has had no significant effect on the Gaussian noise. There is also much less image blurring with this filter in comparison with the median and smoothing filters.

In ImageJ the tool Menu: Process > Filters > Median... applies a median filter using a roughly circular kernel. The radius can be specified by the user. An adaptive filter for removal of outlying pixel values is available as a plugin (the 'Sigma Plus' filter).

## 7.4 Summary

- Image users are mostly interested only in specific parts of the information present in an image. Image filters are used to emphasize, reduce, or remove particular parts of image information or noise in order to more clearly perceive the specific information of interest. Enhancement of edges and attenuation of noise are typical examples of image filtering operations.
- Image filtering can be performed directly on the spatial domain image data, or on the spatial frequency domain data formed by Fourier transformation. The choice of domain depends on the type of filtration required and computational efficiency.
- *Linear* filters have equivalent forms in both domains. This is a direct result of the *convolution theorem*: the process of convolution of an image with a filter kernel in the spatial domain is equivalent to element-wise multiplication of the Fourier transforms of the image and the kernel in the spatial frequency domain.
- An *Ideal filter* is a spatial frequency domain filter with a step-shaped attenuation profile. The amplitudes of all spatial frequencies on one side of the *cutoff* are 100% attenuated, while the amplitudes of all spatial frequencies on the other side of the cutoff are unaffected. An Ideal filter is, in general, not ideal for image processing. It will produce a *Gibbs ringing* artifact parallel to any sharp edges in the image.
- Ringing artifacts can be reduced by use of a spatial frequency domain filter with a gradual attenuation profile. The Butterworth filter is a typical example – the cutoff and slope of its attenuation profile are independently adjustable. In the extreme case a very *high order* Butterworth filter has an attenuation profile that approximates the step profile of an Ideal filter.
- The *Gaussian* filter has a bell-shaped attenuation profile. The slope and cutoff are combined in a single parameter – sigma. A Gaussian filter *never* produces an artifact in the filtered image.
- Specific ranges of spatial frequencies can be filtered with *Band Stop* and *Band Pass* filters. Band stop filters typically have a narrow 'stop band' and are useful for attenuation of specific spatial frequencies, for example periodic noise. Band pass filters usually have a wide 'pass band' and are used to select edge information while simultaneously suppressing noise.
- Filters applied directly to spatial domain image data can be categorized as *convolution* or *non-convolution* filters. Convolution filters (e.g. simple smoothing and edge filters) are linear filters and have equivalent spatial frequency domain filter masks obtainable by Fourier transformation of the

convolution kernel. Non-convolution filters (e.g. median and other rank order filters, and adaptive filters) are non-linear filters and have no spatial frequency domain equivalent.

- The simplest spatial domain convolution filters are characterized by kernels with exclusively non-negative elements. These filters perform a *smoothing* of pixel intensities – the filter blurs the input image. Smoothing filters are called *low pass* filters because their effect is to attenuate high spatial frequencies.
- Edges in images are characterized by pixel *intensity gradients* and can be enhanced or selected with gradient filters. Gradient steepness (equivalent to edge sharpness) can be measured by calculation of the first derivative of the local pixel intensity profile. Maximum gradient steepness (equivalent to edge position) can be measured by calculation of the second derivative of the local pixel intensity profile. Convolution kernels for calculation of gradients include both positive and negative elements. Gradient filters are called *high pass* or *high boost* filters because their effect is to attenuate low spatial frequencies.

# Chapter 8
# Spatial Transformation

## 8.1 Introduction

A spatial transformation of an image is an alteration that changes the image's orientation or 'layout' in the spatial domain. Spatial transformations change the position of intensity information but, at least ideally, do not change the actual information content.

A transformed image might be moved up or down, or left or right, relative to some reference point. Alternatively, it might be rotated, changed in size, or distorted in some way that changes the shape of objects in the image. All these transformations are possible with digital images. Sometimes they require only a simple rearrangement of the positions of pixel data without any recalculation of pixel intensities or colors. In other cases, in attempting to preserve an image's original intensity information and minimize artifacts, it may be necessary to calculate new pixel intensities from the original image data. Even for a well-defined transformation, for example a 10% reduction in size, the method of calculation of new pixel information can have a profound effect on the appearance of the transformed image.

## 8.2 Translation

The simplest image transformation is a spatial translation. A translation might, for example, be used to align anatomy in one image with the same anatomy in another image when the images are overlaid. To perform a translation the image data simply moves relative to some reference point, typically the edge of the image matrix, without changing size or rotating (Fig. 8.1b). To do this it is only necessary to copy the rows or columns of image data into new positions, and to make some decision about what values to put into the newly empty rows and columns. There is no need for calculation of new pixel data and the original data is simply copied to a new position. Apart from the data which has perhaps been moved off the edge of the image matrix there is no change in image quality. This type of operation happens routinely in a computer whenever we use a scroll bar to move the displayed image inside a window.

R. Bourne, *Fundamentals of Digital Imaging in Medicine*,
DOI 10.1007/978-1-84882-087-6_8, © Springer-Verlag London Limited 2010

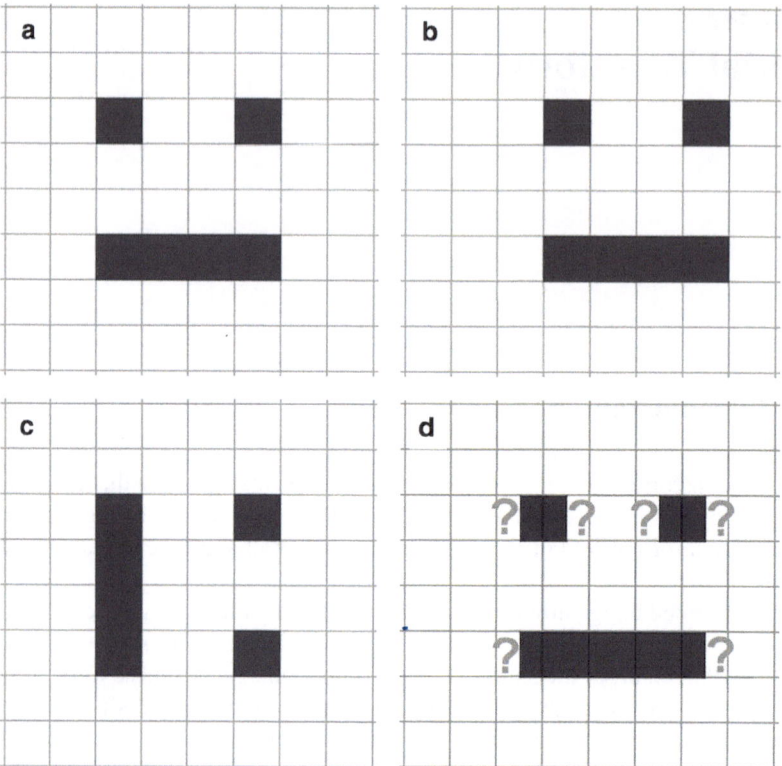

Fig. 8.1 Some simple image transformations. Image **b** shows the data of image **a** shifted (*translated*) one pixel to the right. Image **c** shows the data of **a** rotated *exactly* 90°. Translation is a relatively simple process if the spatial shift is a whole number of pixels – the original image data is simply copied into different positions in the new image matrix. Similarly, rotations that are simple multiples of 90° (**c**) only require a rearrangement of the original image data. A translation that is *not* a whole number of pixels (**d**) is problematic – a decision must be made about how to calculate new pixel values from the original image data. In this illustration the grid represents the rows and columns of the image matrix. In image **d** the 'translated' original image data is illustrated *as if* it was overlaid on the new image matrix. However, this representation has no physical form and the actual pixel values for the new image matrix must somehow be derived from the original data

Rearranging rows and columns of pixels is a relatively simple operation, but complications arises when the required shift is not equal to an integer number of pixels. Exactly what pixel data do we put in the new image matrix when there is no direct correspondence between pixels of the original image and the new image? Imagine that we wish to shift an image half of one pixel to the right (Fig. 8.1d). We might reasonably decide that the best value for each pixel in the new matrix would be the average of the two pixels that 'overly' it in the shifted original image matrix. The simplest way to do this would be to calculate the average of each adjacent pair of pixels on a row of the original image matrix and write this new value into the right pixel position in the new image matrix. This will give us a *representation* of the original image *translated* half a pixel to the right. We will be missing a pixel at the extreme left side of the image, but we will assume this is not a problem. A possibly undesirable side-effect of the averaging method will be a slight blurring of the original image information. If there was a one pixel wide white (intensity 255) line on a black background in the original image it will be converted to a two pixel wide mid-gray (intensity 128) line in the translated image. In a similar way all vertical edges in the original image will become less sharp. Because we shifted our image horizontally, and thus only averaged pairs of pixels along rows, there will be no blurring in the vertical direction – horizontal edges will retain their original sharpness. The averaging of pairs of pixels in the image has caused a loss of some of the original intensity information. Obviously there are potentially serious problems arising from a pixel-averaging approach to the image translation problem. Some better methods are described below.

## 8.3 Rotation

A 90°, 180°, or 270° rotation of an image does not require any recalculation of pixel data. For example, to perform a 90° rotation all that is necessary is to copy all the rows of the original image data into the appropriate columns of a new image matrix (Fig. 8.1c). An $m \times n$ pixel image (rows × columns) will become an $n \times m$ pixel image. Similarly, flipping an image horizontally or vertically is only a matter of data rearrangement and requires no calculation of pixel data.

Rotating an image by an angle that is not a multiple of 90 degrees is a less trivial process because the rotated original image data no longer align exactly, pixel by pixel, with the new image matrix (Fig. 8.2). What values do we put in the new image matrix? Figure 8.3 illustrates two simple solutions to this problem based on how

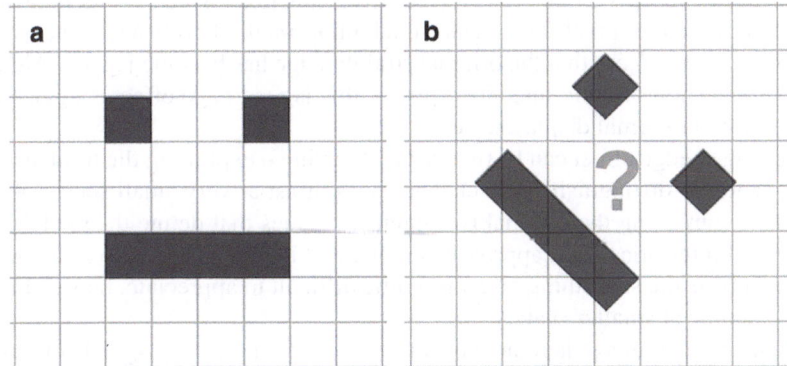

**Fig. 8.2** When an image is rotated by an angle that is not a multiple of 90° the original pixel data cannot be mapped exactly, pixel by pixel, to the new image matrix. Figure 8.3 illustrates two trivial, but unsatisfactory, solutions

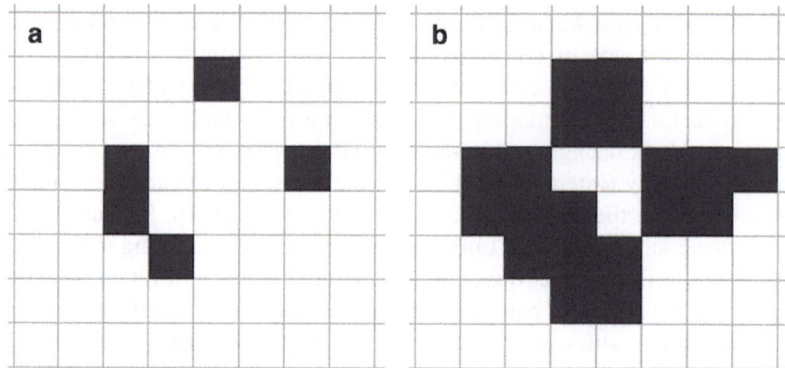

**Fig. 8.3** Two possible (but rather unhappy looking) solutions to the image rotation problem posed in Fig. 8.2. In image **a** any pixel in the new image matrix that is overlaid by *more than half* a dark pixel of the rotated original image is given the dark value. In image **b** any pixel in the new image matrix that is overlaid by *any fraction* of a dark pixel of the rotated original image is given the dark value

much overlap there is between rotated original pixels and new image pixels. One approach is to make all new matrix pixels dark if any part of an original dark pixel overlies it. The second approach only makes a new image pixel dark if more than half an original dark pixel overlies it. In this example the original image information is very badly distorted by both the rotation processes – one method gives an image with too few dark pixels and the other method gives too many.

We can intuitively see that a better approximation to the original image appearance is possible, but we need some sort of rule or algorithm to calculate the new pixel values. Perhaps we should aim to have about the same number of dark pixels in our rotated image as in the original. One way to achieve this would be to give each pixel in the new image matrix the value of the *nearest* pixel in the rotated matrix. What is the nearest pixel? Let's say it is the overlying pixel whose center is closest to the center of the new pixel. This 'nearest neighbor' method gives a reasonably good result (Fig. 8.4a) except that the original straight edge has become jagged. Although this example shows highly magnified pixels, this jagged edge effect ('jaggies') can often be seen at normal display scales.

The jagged edge effect can be reduced by blurring – exploiting the limited ability of our vision to distinguish small changes of contrast at very small scales. We can do this by converting the original two intensity values that define the straight edge into several intensities. This approach is illustrated in Fig. 8.4b. In the example with enlarged pixels the smoothing effect is a little difficult to appreciate, but see Fig. 8.5 for an example at smaller scale.

When an $m \times n$ image is rotated by some arbitrary angle the original corners will not always lie inside an $m \times n$ or $n \times m$ matrix. If the information in the corners of the original image is important then the new image matrix will have to be made larger than the original and the 'empty' corners filled with some appropriate background values.

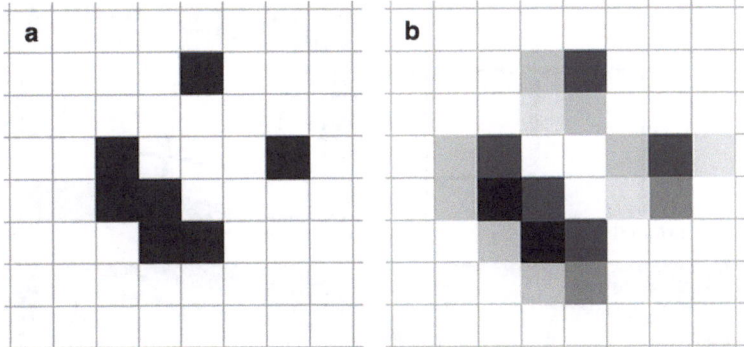

**Fig. 8.4** More sophisticated solutions to the image rotation problem posed in Fig. 8.2. In image **a** pixels acquire the value of the original pixel whose *rotated* center position lies closest to the center of the new matrix pixel. In image **b** pixels acquire a value that is the product of the rotated original image pixel value and the degree to which it overlies the new pixel. These methods provide visually better representations of the original image data than the simpler methods illustrated in Fig. 8.3

The problems of how to represent a rotated edge, or a half pixel translation, are part of the more general problem of how to calculate pixel values for a transformed image when there is no direct correspondence between the transformed original pixels and the new image matrix. The general name for the process is *interpolation*.

## 8.4 Interpolation

Interpolation is the calculation of the expected value of a function at a particular point when only the values at nearby points are known. In the case of 2D images we are interested in 2D functions that describe image intensity relative to two spatial coordinates. There are three common interpolation methods used in 2D image processing: *Nearest-neighbor*, *Bilinear*, and *Bicubic*.

### 8.4.1 Nearest-Neighbor

We introduced the 'nearest-neighbor' idea in the image rotation section above, but it can be applied to all image transformations. This method calculates the distance between the 'notional' centers of the transformed image matrix pixels and the centers of the new matrix pixels. Each pixel in the new image matrix is assigned the *exact* value of the nearest pixel in the rotated original matrix. Because there is no adjustment of pixel intensities this method preserves contrast differences between adjacent pixels. Sharp edges are not blurred but they may appear spatially distorted at the pixel level – either becoming jagged if they were originally vertical or hor-

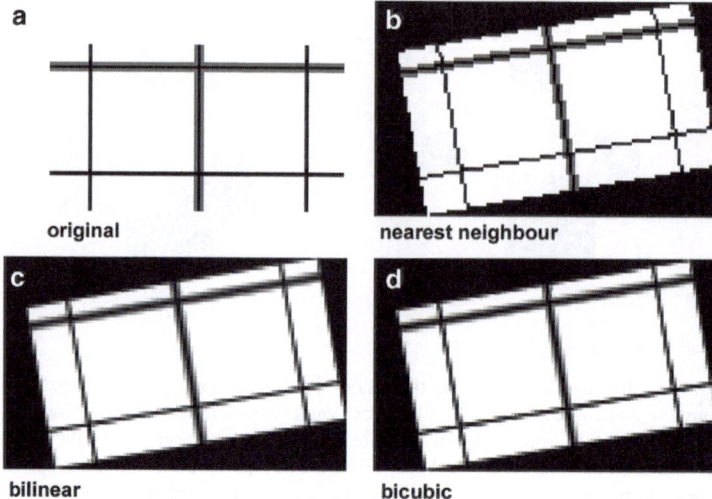

a        original

b        nearest neighbour

c        bilinear

d        bicubic

**Fig. 8.5** Comparison of three different interpolation methods applied to the problem of image rotation. The original image (**a**) includes both sharp and blurred edge detail. Nearest neighbor interpolation (**b**) preserves the original sharp edge detail but produces conspicuous jagged aliasing artifacts. Bilinear interpolation (**c**) reduces aliasing but blurs the sharp edges. Bicubic interpolation (**d**), for this particular image, gives a result intermediate between nearest neighbor and bilinear. These example images have been enlarged to illustrate the pixel-level effects, some of which may not be obvious at more conventional display scales (see Fig. 8.6). Note that the rotated image requires a larger matrix than the original. Here the background has been filled with black pixels

izontal, or possibly becoming smooth and straight if they were originally jagged. These distortions are called *aliasing*.

## 8.4.2   Bilinear

The bilinear method calculates a new pixel value from the four centers of the transformed original pixels that surround the new pixel center. The new pixel value is calculated by linear interpolation in both the $x$ and $y$ directions – hence the term 'bilinear'. The value for the new pixel is the average of the nearest four original pixels *weighted* according to the proximity of the new pixel center to the centers of the four original pixels. Bilinear interpolation leads to some blurring of edge detail but, for the same reason, it reduces the creation of jagged aliasing artifacts.

## 8.4.3   Bicubic

The bicubic method is a more sophisticated version of the bilinear method. Instead of connecting pairs of points with straight lines the bicubic method fits two second order polynomials (hence 'bicubic') to the *sixteen* pixels of the transformed original image matrix that are nearest to the center of each new image pixel. Because the bicubic method can represent a 'curvature' in the intensity profile of the original image it produces transformed images that more closely emulate the original than the bilinear method – especially if the image does not contain very sharp edges and lines.

Comparisons of the three interpolation methods are illustrated for a simple test image (Fig. 8.5) and a detailed medical image (Fig. 8.6). It is plain that for anatomical medical images viewed at normal scale these three different interpolation methods do not produce markedly different effects. The general consensus from photographic image processing is that bicubic interpolation gives the best looking

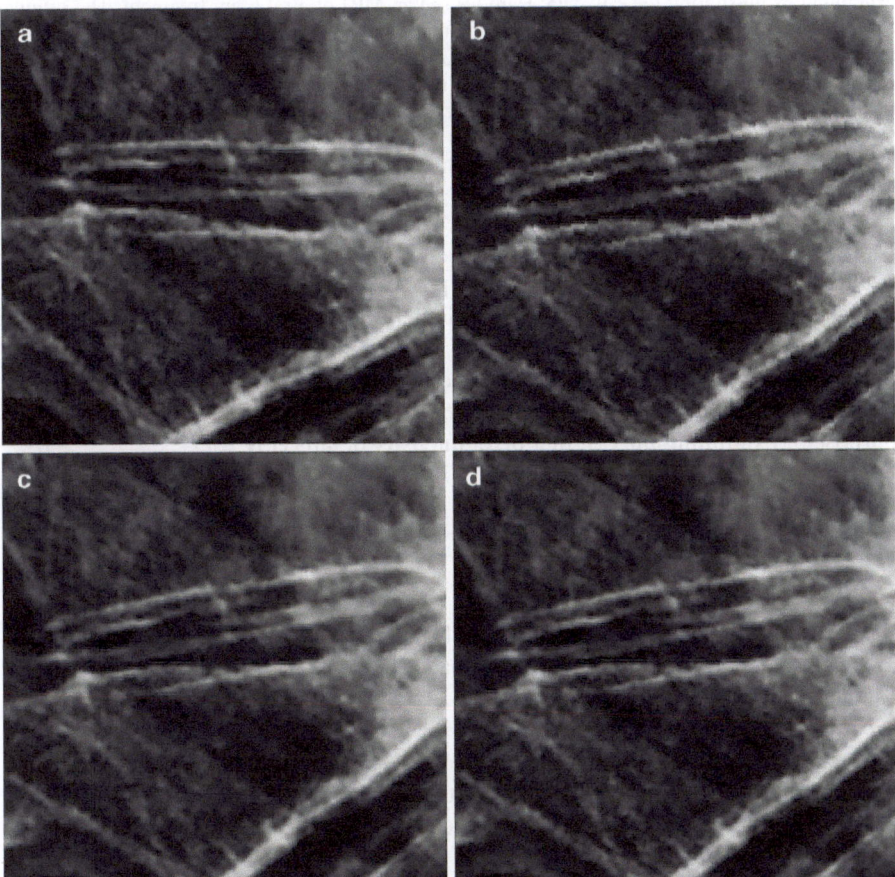

**Fig. 8.6** Comparison of interpolation methods applied to rotation (10°) of a detailed X-ray image. The original image (**a**) was 200 × 200 pixels. The nearest neighbor method (**b**) produces just-detectable aliasing artifacts along the sharpest edges. The bilinear (**c**) and bicubic (**d**) methods appear artifact-free when the image is viewed at a conventional scale

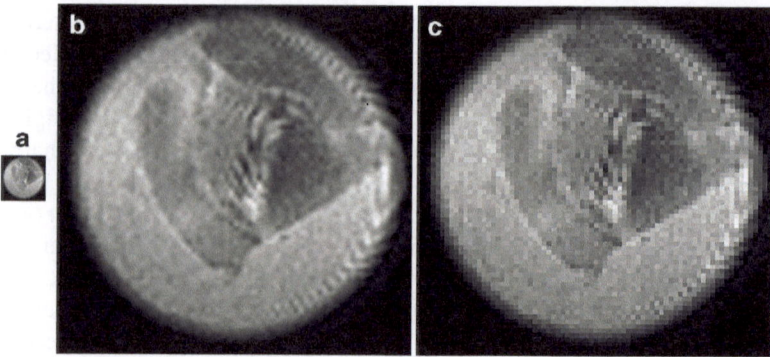

**Fig. 8.7** Nearest neighbor interpolation is useful when enlarging an image for the purpose of examining or illustrating the raw data. Here we see an MRI of a rat brain fragment. The original $80 \times 80$ pixel image (**a**) is much too small to study when displayed (in this case printed) as an $80 \times 80$ pixel print image. Because the printer works at about 118 dots/cm we need to increase the original image matrix to about $590 \times 590$ pixels to make the print size $5\,cm^2$. Enlargement by the bicubic method (**b**) blurs the boundaries between the original pixels. The nearest neighbor method (**c**) preserves the original pixel information. For the original pixel information to be completely undistorted the enlargement factor must be an integer (in this image the unusual edge pattern is an MRI artifact not related to image enlargement)

results for most images, and most image processing software uses bicubic interpolation as the default method. The extra calculations needed for bicubic interpolation are not a significant problem for modern computers. However, this does not mean that bicubic interpolation should be used for *all* image transformations. Nearest neighbor interpolation is useful when we want to enlarge images for the specific purpose of examining the original pixel intensity details (Fig 8.7).

## 8.5 Resizing Images

We resize images on monitor displays so frequently that we rarely give any thought to the process, but changing the size of images presents the same sort of problems in calculation of the new image pixel values as does translation and rotation (Fig. 8.8). Most of the time the bicubic method is ideal – it is built in to the graphics card of computers so that the displayed image size can be changed easily without obvious artifacts, and it is the usual default option for resizing image data in image processing software. The smoothing effect of bicubic interpolation can be used to suppress pixelation effects.

Sometimes the spatial resolution of raw image data is such that pixelation is clearly visible if the image is viewed at its native resolution. This is especially the case for imaging techniques that have poor spatial resolution, or small image matrices – e.g. some MRI methods. The propensity of human visual perception to notice lines and edges means that pixelated images are generally regarded as difficult to

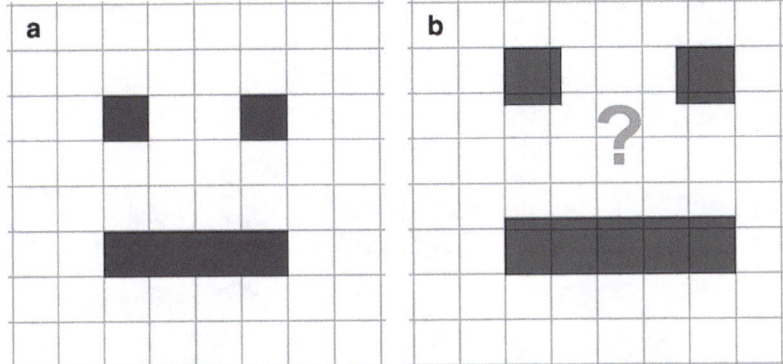

**Fig. 8.8** When an image is scaled, in this example enlarged by 25%, similar problems occur to those we saw for rotation in Fig. 8.2 – the original pixel data cannot be mapped exactly, pixel by pixel, to the new image matrix

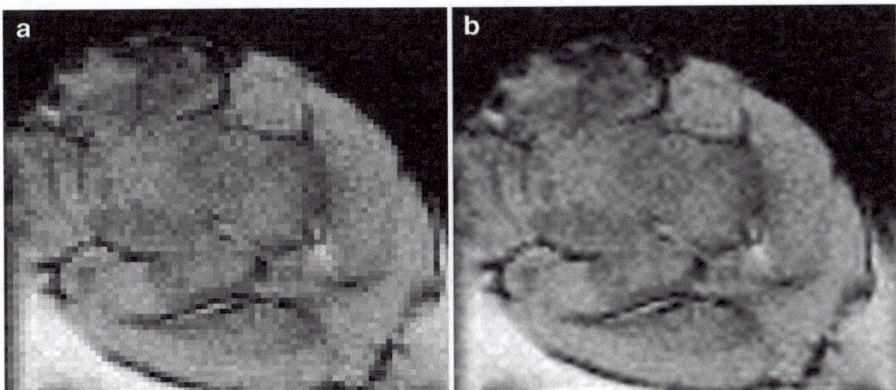

**Fig. 8.9** Intentional blurring of pixel boundaries can enhance the 'readability' of anatomical information. This mouse brain was originally imaged with a 128 × 128 matrix – part of which is illustrated as image **a**. Bicubic interpolation (image **b**) increases only the *apparent* spatial resolution and blurs the raw data pixel boundaries. The anatomical information is easier to see in image **b**. In Fig. 8.7c we chose the nearest neighbor method to enlarge and illustrate the raw pixel data. Here the printed 'raw data' image (**a**) is also the result of nearest neighbor interpolation

interpret because the visible edges of the pixels distract from the tonal information. These images are often interpolated into a larger image matrix which, when displayed, does not exhibit obvious pixelation. A typical example is the MRI data illustrated in Fig. 8.9 which shows both the raw and interpolated image data. Interpolation increases the *apparent* spatial resolution.

It is important to remember that interpolation of spatial information in order to increase the apparent spatial resolution and 'readability' of an image does not add information to the image. The spatial resolution of an image is absolutely limited by the spatial resolution of the raw data.

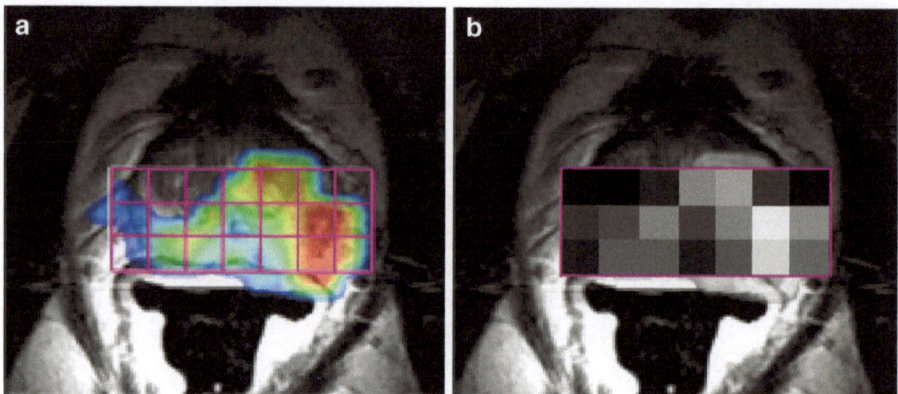

**Fig. 8.10** Interpolation of magnetic resonance spectroscopy data. High resolution 'metabolite maps' (**a**) are commonly derived from low spatial resolution raw data (**b**) acquired from a specific measurement region – inside the purple rectangle. The high resolution map is easier to interpret than the raw data but suggests spatial details, both within and outside the measurement region, that have not been measured. The true spatial resolution is in fact lower than the nominal resolution (individual purple boxes) due to the point spread function

Spatial interpolation is very commonly used to aid in the interpretation of low spatial resolution data, but it can easily lead to interpretation errors. Figure 8.10 illustrates a typical color 'metabolite map' derived from magnetic resonance spectroscopy (MRS) of the human prostate. The nominal raw spatial resolution of the data is of the order of 0.5–1 cubic centimeters (Fig. 8.10b). This low-resolution data is difficult to interpret so it is interpolated into a high resolution color contour map overlaid on the anatomical image (Fig. 8.10a). This contour map *suggests* a level of spatial detail that is not present in the acquired data, and even includes contours in regions from which no measurements were acquired. If the raw data is not examined carefully there is a strong tendency for the viewer to forget that the true spatial resolution of the measurement is very coarse and to overinterpret the high resolution image.

## 8.6  Summary

- Image transformation – the enlarging, shrinking, or rotating of an image – requires generation of new image matrices. In general, the pixel intensities in the transformed image matrix must be calculated from those in the original image matrix. Pixel intensities in the transformed image are calculated or assigned by interpolation methods – most commonly *nearest neighbor* and *bicubic*.
- The nearest neighbor interpolation method assigns only original image matrix intensity values to pixels in the transformed image matrix. The value

assigned is the value of the nearest pixel center in the notionally transformed original image matrix. The nearest neighbor method may produce 'jagged' edge artifacts, but is useful for display of original intensity information in transformed images.

- The bicubic interpolation method calculates new intensity values for pixels in the transformed image matrix. The value assigned is the estimated local value on a curved surface representing the intensities of the notionally transformed original image matrix. The bicubic method is often used to produce more visually appealing images by suppression of pixelation, however, this may lead to misinterpretation of spatial resolution. Interpolation *cannot* create new image information.

# Appendix A
# ImageJ

This appendix briefly outlines where you can get ImageJ and find information on how to operate and customize it. ImageJ is under constant development with many users developing plugins for specific tasks and making them publicly available. Read the *News* page on the ImageJ website to check the progress of modifications to the base version. Read the *Features* page for a detailed listing of the current functionality.

## A.1  General

ImageJ is a public-domain, Java-based image processing program developed by Wayne Rasband at the National Institutes of Health, USA. It is free to download from the ImageJ website (http://rsb.info.nih.gov/ij/index.html) and will run on any computer.

Figure A.1 shows a sample screen view of the ImageJ user interface. Here we see the main menu bar at the top and four image windows. In the example the windows are (clockwise from top left): (1) An MR image of an orange; (2) The Fourier spectrum of the orange image; (3) A band pass filter created with the band pass filter tool; and (4) the image of the orange after frequency domain processing with the band pass filter.

The image windows, any windows displaying the output of processes or measurements (e.g. a histogram), and the control windows for the tools all float independently on the computer desktop rather than being contained inside a main application window. If you find the lack of a plain background distracting open a simple application such as Notepad and maximize its window. You can use this empty window as a plain background.

R. Bourne, *Fundamentals of Digital Imaging in Medicine*,
DOI 10.1007/978-1-84882-087-6_9, © Springer-Verlag London Limited 2010

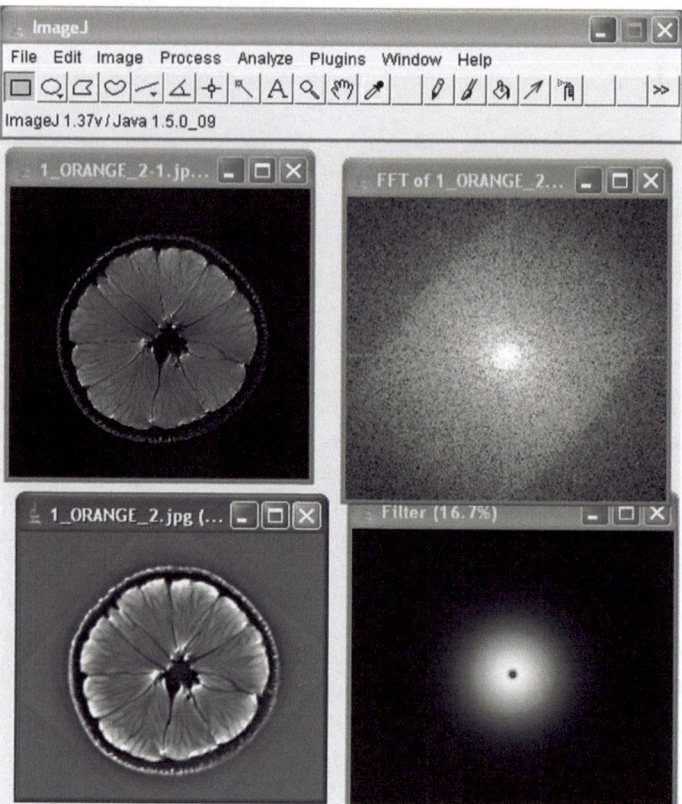

**Fig. A.1** A typical ImageJ screenshot

## A.1.1 Installation of ImageJ

Download the base version of ImageJ appropriate for your computer (Windows, Mac, or Linux). The safest way to be sure the program will run on your computer is to select the download which includes the Java runtime environment. Follow the installation directions on the Download pages. Most of the image processing tasks described in this book can be performed with tools included in the base version of ImageJ.

## A.1.2 Documentation

There are several sources of introductory, instructional, and reference documentation on the ImageJ website. These can also be accessed via *Help* on the ImageJ menu bar. If you use ImageJ a lot you will inevitably discover a few idiosyncrasies and bugs. There is a very active and helpful mailing list where advice can be sought. Subscribe via the ImageJ website.

### A.1.3  Plugins

The open architecture of ImageJ means its functionality can be extended by third party plugins and recordable macros. If you cant find the tool you want, such as a particular type of filter, in your installation of ImageJ there is a good chance you will find it available as a plugin. The first place to look is on the ImageJ plugins page (http://rsb.info.nih.gov/ij/plugins/index.html).

## A.2  Getting Started

Start by spending 15 min reading the *Introduction, Basic Concepts* and *Overview* sections from the ImageJ *Documentation* page. This will give you a good feel for the way ImageJ is used.

## A.3  Basic Image Operations

Read the *Menu Commands* and *Tools* sections on the Documentation page. The simplest image operations are measurements of image characteristics. You can open some images of your own, or one of the sample images automatically installed with ImageJ (Menu: File > Open Samples). Some filtering operations produce outputs with negative pixel values. Negative pixel values will be converted to zeros in most image formats. In order to visualize negative values first convert the image to 32-bit mode before processing.

## A.4  Installing Macro Plugins

Installing a plugin is simply a matter of copying the downloaded files into the ImageJ Plugins folder on your computer (usually C:\Program Files\ImageJ\Plugins on a PC) and restarting ImageJ. The installed plugin should then appear in the *Plugins* menu.

## A.5   Further Reading

As well as the basic documentation the ImageJ website has links to documentation describing writing macros and plugins, and a Wiki reference. For an extended coverage of the applications of ImageJ, including Java code, refer to the textbook *Digital Image Processing: An Algorithmic Introduction Using Java* by Wilhelm Burger and Mark Burge (Springer-Verlag, 2007).

# Appendix B
# A Note on Precision and Accuracy

The terms *precision* and *accuracy* are often used interchangeably but they have distinctly different meanings when used correctly. Precision refers to the repeatability of a measurement, accuracy to the correctness of the measurement.

Here's an example: Imagine weighing a coin on a laboratory balance. The display says 5.002 g. Is this measurement accurate? Is it precise? You remove the coin from the balance, check that the display returns to 0.000, and then weigh the coin again. This time the display says 5.005 g. Which measurement do you believe to be correct? Just to be sure, you repeat the process three more times, each time checking that the display returns to 0.000, and get readings of 5.000, 5.004, and 5.001 g. The mean and standard deviation of the measurements are 5.002 g and 0.002 g respectively.

Now you notice that the balance has a *Calibrate* button which you press (after removing the coin). After calibration you repeat the five measurements with results of 4.903, 4.906, 4.901, 4.905, and 4.902 g. This time the mean and standard deviation of the measurements are 4.903 and 0.002 g, respectively.

Notice that in both sets of measurements the standard deviation of the measurements was identical. However, before calibration the average of the measurements was too high by 0.099 g. The precision was the same ($\pm 0.002$) for each set of measurements but the accuracy depended critically on calibration. If the balance had not been properly maintained then we might expect some friction problems in the mechanism to lead to a greater standard deviation in the measurements, or a *lower precision*.

Beware, it is easy to be deceived by the number of digits on a display that a measurement is both precise and accurate when neither may be the case. Similarly, we need to be careful about specifying measurements with meaningless decimal places. If the standard deviation of the above measurements of the coin's weight was 0.02 g rather than 0.002 g, then it would be inappropriate to describe the weight of the coin with milligram precision because the uncertainty is at least 20 mg.

Note that if we wanted to digitally encode our measurements above we would need, at the very least, 13 bits ($2^{13} = 8192$) to store the measurements with adequate precision, since we want to measure 5,000 mg in increments of 1 mg.

R. Bourne, *Fundamentals of Digital Imaging in Medicine*,
DOI 10.1007/978-1-84882-087-6_10, © Springer-Verlag London Limited 2010

# Appendix C
# Complex Numbers

Since we use complex numbers in several aspects of image processing and analysis introduced in this text it is important to have a basic grasp of what they are and how they are used. This appendix gives a very basic outline for readers who have not previously encountered them.

## C.1  What Is a Complex Number?

We can define a complex number $Z$ as a number of the form: $Z = a + bi$ where $a$ and $b$ are real numbers and $i = \sqrt{-1}$. For the purposes of this discussion a real number is a number that can be written as a decimal. The product of a real number and $i = \sqrt{-1}$, the $bi$ part of $Z$ above, is called an *imaginary number*. So a complex number has a *real* part and an *imaginary* part.

Although $\sqrt{-1}$ appears to have no intuitive meaning in our everyday world, it is an extraordinarily useful concept in the science and mathematics. It can be used to describe familiar phenomena like the flow of electricity, and not-so-familiar things like the behavior of individual electrons.

## C.2  Manipulating Complex Numbers

It is often useful to depict complex numbers as points on a *complex plane* as shown in Fig. C.1. In the complex plane all the real numbers lie on the 'X' axis, labelled 'R' for Real in this diagram. The imaginary numbers lie on the vertical 'Y' axis, here labeled 'I'.

A line, or vector, drawn from the point representing zero (the origin) to the point representing $Z$ makes an angle $\theta$ with the real axis. The length of the line from the origin to point $Z$, the *magnitude* of $Z$, is $|Z| = \sqrt{a^2 + b^2}$. Note that $i$ is not included in the magnitude calculation. The magnitude can *only* be positive or zero. We can also express $a$ and $b$ in terms of $|Z|$ and $\theta$:

$$a = |Z|cos(\theta), \text{ and } b = |Z|sin(\theta).$$

R. Bourne, *Fundamentals of Digital Imaging in Medicine*,
DOI 10.1007/978-1-84882-087-6_11, © Springer-Verlag London Limited 2010

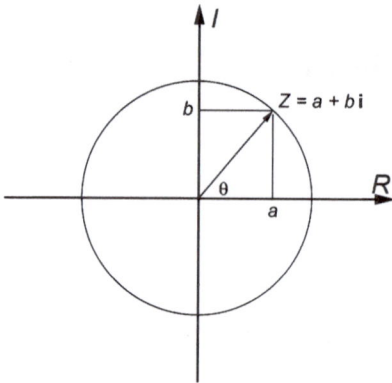

**Fig. C.1** The complex number plane. The complex number $Z$ has a real component $a$ and an imaginary component $bi$. In the complex plane all the real numbers lie on the 'X' axis, labelled 'R' for Real in this diagram. The imaginary numbers, multiples of $i = \sqrt{-1}$, lie on the vertical 'Y' axis, here labeled 'I'. The circle describes all the complex numbers that have magnitude $|Z|$

The point representing i lies on the imaginary axis one unit length above the origin. $i^2 = -1$ lies on the real axis one unit length to the left of the origin, $i^3 = -i$ lies on the imaginary axis one unit length below the origin, and $i^4 = 1$ lies on the real axis one unit length to the right of the origin.

Perhaps you notice a pattern here? Multiplying complex numbers is the same as adding the angles that lines drawn to them from the origin make with the positive real axis, and multiplying their magnitudes.

Manipulating complex numbers is often much easier if we use Euler's Formula:

$$e^{ix} = cos(x) + isin(x) \tag{C.1}$$

Now imagine we want to multiply two complex numbers $Z_1$ and $Z_2$:

$$Z_1 \times Z_2 = |Z_1|e^{i\theta_1} \times |Z_2|e^{i\theta_2} = |Z_1||Z_2|e^{i(\theta_1+\theta_2)} \tag{C.2}$$

Or if we are dividing,

$$\frac{Z_1}{Z_2} = \frac{|Z_1|e^{i\theta_1}}{|Z_2|e^{i\theta_2}} = \frac{|Z_1|}{|Z_2|}e^{i(\theta_1-\theta_2)} \tag{C.3}$$

In the complex plane we measure angles in *radians*, not degrees. There are exactly $2\pi$ radians in a full circle. Remember that the circumference of a circle of radius $r$ is $C = 2\pi r$, so one radian is simply the angle made by tracing a distance of one radius around the circumference of a circle. The use of radians rather than degrees to measure angles greatly simplifies complex number calculations.

If we substitute $x = \pi$ in Euler's formula we get:

$$e^{i\pi} + 1 = 0 \qquad\qquad (C.4)$$

– an expression considered by some to be the most beautiful theorem in mathematics because it interrelates $\pi$, $e$, i, 1 and zero in a single simple expression. Richard Feynman called it *'Our shining jewel'*.

When we want to display an image *representing* a 2D matrix of complex data, such as the Fourier spectrum of a digital image, or raw MRI data, we usually display the magnitude data for each matrix element (image pixel). This gets us around the problem of how to show the real and imaginary parts of the data and how to show negative components. Because the magnitude of the zero frequency component of the Fourier spectrum of a digital image is usually much greater than the magnitude of the non-zero frequencies, we usually plot the log of the magnitudes.

What is the relevance of all this to measurement and representation of frequencies? To answer this question we need to think about a vector $\mathbf{Z}$ that *rotates* around the origin in the complex plane.

Let's imagine that $|Z| = 1$ and that $\mathbf{Z}$ is rotating steadily anticlockwise around the circle shown in the diagram, making one complete rotation of the circle every second. As $\mathbf{Z}$ rotates we will record the values of $a$ and $b$. Since $|Z| = 1$ we find simply that $a = cos(\theta)$ and $b = sin(\theta)$. Our recordings of $a$ and $b$ are simply sine waves, varying continuously from $+1$ to $-1$, with frequency one cycle per second. The sine wave for $a$ is 90 degrees *out of phase* with the sine wave for $b$, in other words, when $a = \pm1, b = 0$ and vice-versa.

Here are some direct physical applications of complex numbers:

## C.3  Alternating Currents

Almost all modern electricity is produced by *three phase* generators. Three huge coils of wire are connected in a 'Y' pattern and spin on an armature inside an electromagnet. The voltages generated in the three coils can be described by a complex plane diagram (Fig. C.2). Now we have three vectors $\mathbf{V_1}$, $\mathbf{V_2}$, $\mathbf{V_3}$ originating from the center and displaced 120 degrees from each other. If the generator armature spins at 3,000 rpm (50 Hz) then we can think of the three vectors spinning around the origin at 3,000 rpm (or $100\pi$ radians . $s^{-1}$). If we measure the voltage $v_1$ generated in the coil represented by $\mathbf{V_1}$, relative to the center of the 'Y', we observe a sine wave. The voltage $v_1$ is the *real* (x-axis) component of $\mathbf{V_1}$. We can *only* measure the real component. If we measured $v_2$ and $v_3$ we would find they also vary sinusoidally and have the same amplitude $|V|$ as $v_1$, however, $v_2$ and $v_3$ are 120° *out of phase* with $v_1$.

What would we observe if we were to measure the voltage between the ends of two of the generator coils, let's say those represented by $\mathbf{V_2}$ and $\mathbf{V_3}$? We can answer this without even making the measurement. We simply subtract $\mathbf{V_2}$ from $\mathbf{V_3}$, as shown by the gray arrows in Fig. C.2. Now we have a new arrow $\mathbf{V_2}$–$\mathbf{V_3}$ that rotates at the same speed as $\mathbf{V_2}$ and $\mathbf{V_3}$. If we do a little bit of trigonometry we find that the amplitude of this voltage is 1.73 times $|V|$ and its sinusoid is 30° out of phase with $v_2$.

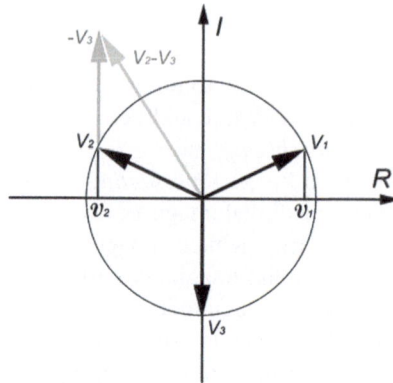

**Fig. C.2** Complex plane representation of the relationship between the voltages produced by a three phase generator

## C.4 MRI

A similar diagram is useful for describing how we measure an MRI signal. We observe the tiny oscillating voltage induced in a radio frequency coil by the net magnetization of hydrogen nuclei precessing in a magnetic field. A single coil can only measure magnetization in one plane. If multiple detector coils are used we can get more signal (relative to noise) and can combine the measured signals after appropriate corrections for the phase differences. We would not be measuring an *imaginary* voltage in *any* of the coils. If we rotate the plane of a coil then, effectively, we rotate the real axis of the voltage diagram for that coil. The voltages we measure are real. It is the mathematical formalism that requires the imaginary numbers, but it also provides us with the information on how to make the phase corrections.

In MRI the raw data is always complex. How does this fit with our statement that we can only measure real voltages? Basically, we *synthesize* the imaginary data. If we have two detector coils arranged at right angles to each other then we can store the voltages measured by one of the coils in the imaginary part of the raw data matrix and the voltages measured by the other coil in the real part.

The magnitude of the measured voltage tells us (roughly) how many hydrogen nuclei are contributing to the signal. The frequency of the measured oscillation and the phase relative to some reference tell us the origin of the signal in space, i.e. the anatomical location. With this information we can construct a 2D map of signal intensity versus spatial location – an image of the anatomy. All we need to do to make this image is an inverse 2D Fourier transformation of the complex raw data.

The above examples illustrate the application of complex numbers to describe and predict physical processes that have a conspicuous periodic, or oscillating, nature. We can also use complex numbers in a more artificial sense, for example in the mathematical description of the shape of objects in images. In this case we take a completely static entity – the outline of an object, and *pretend* that it lies in the complex plane. Because the list of boundary points eventually comes back to the same place we can think of the list as being a discrete sampling of a periodic function. We then perform a 1D Fourier transform on the list of the boundary points. The resultant *Fourier Descriptor* gives a neat summary of the shape's morphological properties and permits a quantitative comparison of different shapes.

# Index